EUPHRATES AND TIGRIS
MESOPOTAMIAN ECOLOGY AND DESTINY

MONOGRAPHIAE BIOLOGICAE

Editor

J. ILLIES

Schlitz

VOLUME   38

Dr. W. Junk bv Publishers   The Hague – Boston – London   1980

EUPHRATES AND TIGRIS, MESOPOTAMIAN ECOLOGY AND DESTINY

JULIAN RZÓSKA

with contributions by

J. F. Talling, F.R.S. and Dr. K. E. Banister

Dr. W. Junk bv Publishers   The Hague – Boston – London   1980

Distributors:

for the United States and Canada

Kluwer Boston, Inc.
160 Old Derby Street
Hingham, MA 02043
USA

for all other countries

Kluwer Academic Publishers Group
Distribution Center
P.O. Box 322
3300 AH Dordrecht
The Netherlands

Library of Congress Cataloging in Publication Data   CIP

Rzóska, Julian.
  Euphrates and Tigris, Mesopotamian ecology and destiny.
  (Monographiae biologicae; 38)
  Bibliography: p.
  Includes index.
  1. Ecology—Iraq. 2. Aquatic biology—Iraq.
3. Man—Influence on nature—Iraq. 4. Iraq.
I. Talling, John Francis, joint author. II. Banister, Keith Edward, joint
author. III. Title. IV. Series.
QP1.P37 [QH193.I7] 574s [574.5'09567'4] 80–435

ISBN-13: 978-94-009-9173-6   e-ISBN-13: 978-94-009-9171-2
DOI: 10.1007/978-94-009-9171-2

... Assyria, a country remarkable for the number of great cities ...

The rainfall of Assyria is slight and provides enough moisture to burst the grain and start the root growing ... but ... to bring the grain to maturity artificial irrigation is used. ...

Herodotus, 4th century B.C.,
Histories, Book I.

Contents

Remark. Chapters 6, 6a and 8 have their own references, as they will be available as reprints.

Introduction
Scope and limitations of this book

I am trying here to present the natural history of a land largely created and dominated by two great rivers, the Euphrates and Tigris. All rivers have two main functions, quite different from lakes; they transport water and eroded material sometimes over large distances. The astute Greeks, who penetrated here in the 4th century B.C., called the land Mesopotamia, an apt name; it is the only region in the Near East, except Egypt, having the benefit of large rivers. Another name coined in antiquity was 'Fertile Crescent', stretching from Egypt to present day Iraq; Herodotus marvelled at the fertility of the soils, the abundance of water and the magnificent cities of Mesopotamia. Thus a further role of some great rivers is recognized as foci of human development. The desire to collate this book arose from a similar motif as in the Nile book (1976), the intricate connection between man and rivers.

Such a motif can only be realized by an interdisciplinary approach with many difficulties involved; geological events create rivers, climate governs the water supply, the surrounding land influences the vegetation and the physical and chemical features of water. The great complexity of any land, examined on ground level, has been done in our case of Mesopotamia from, so to speak, a 'birds eye' view. The space photographs used even enlarge this way of looking at the country from a distance. This omits many details but brings out the essential features; text and illustrations are integrated.

The term 'Mesopotamia' is used although most of the book is devoted to the present political entity of Iraq, but natural history is not confined to political areas, whose problems of political implications are not my task. The book is divided into two parts, the first deals with the land, its geology, morphology, climate and vegetation, its past and man's emergence as dominating agent. This gives the general setting for the second part, devoted to the waters of present Iraq and of the wider ecological panorama of the Near East. Only against this background can the role of the two rivers, the Euphrates and Tigris with their tributaries, be assessed properly. The hydrological regime of the river system is strongly marked by its instability which has been recorded from ancient times onwards. Spring floods cause changes in the regime with wide inundations, changes in the courses; at low water the force of the rivers dissipate forming marshes. Irrigation practised for the last 6000 years has caused salinity of large parts of the alluvial soils. All this is reflected in the physical, chemical and biological features of the waters. Many waters of the Near East share the salination problem, inherent to arid regions, some of these in regions adjacent to Iraq have been discussed. I am aware of the short-comings of limnological documentation, based solely on local, often insufficient sources. Thus this book is only a framework to be filled in years to come by local scientists. Two co-authors have helped by writing special chapters: Dr. J. F. Talling, F.R.S. has written on water characteristics and phytoplankton, Dr. K. E. Banister (B.Mus. Nat. Hist.) has critically summarized our present knowledge on the fishes.

An *Epilogue* summarizes the main results, intended for those readers and reviewers who have no time for reading the whole book; some general and personal remarks are added. An *Annexe* contains the impressions of eyewitnesses of the Mesopotamian scene in the past.

ACKNOWLEDGEMENTS. To my two co-authors I owe gratitude for helping expertly in difficult problems. Dr. J. F. Talling has read the whole script and prevented me from making errors both factual and stylistic. Mr. Guest (Roy. Bot. Gardens, Kew) has supplied invaluable information, gained from his long experiences in Iraq; Dr. Falcon F.R.S. drew my attention to the formation of the delta; Dr. Macan has informed me on the malaria problem in Iraq and the vectors from rich personal experience. Dr. Colette Serruya kindly gave me the information on the Dead Sea. Dr. Macan has provided me with the volume of the 'Handbook of Iraq and the Persian Gulf' issued by Naval Intelligence during the last war. This book is still the most comprehensive source of information on that region, though published in 1944.

For permission to use illustrations from this book I owe gratitude to the Ministry of Defence, London, succeeding Naval Intelligence. The Iraq Petroleum Company, now an independent body, gave me some excellent photographs; Mr. Guest raised no objections against using his graphs and maps for my adaptations. Dr. Talling used some diagrams published previously in the 'Internationale Revued. gesammten Hydrobiologie': space photographs were obtained commercially from the EROS Centre in the U.S.A.

The rendering of Arabic names is unequal.

August 1979   Julian Rzóska
                              6 Blakesley Avenue, London W.5.

Part I

The lands and their life

1 Panorama of Mesopotamian Iraq

Geology, climate, soils

Iraq is only part of a great area of arid lands, stretching from North Africa through the Near East, Iran to Beluchistan and Western Asia. There are some similar features of this vast area but also considerable differences in the main factors governing the nature of any land, that is geomorphology, climate, soils and water.

Geology

Only a simplified account is necessary in the context of this book. Three main geological regions exist in Mesopotamian Iraq and beyond. The Arabian shield in the west and south is the most extensive region and covers 57 per cent of Iraq and vast areas beyond. Secondly, the foothills and mountains in the east and north occupy 18 per cent of the Iraq territory; the third region is the alluvial plain, bordered by the two previously mentioned regions; it occupies 25 per cent of the country.

The *Arabian shield* is an old, undisturbed formation occupying parts of north Africa, Sinai, Arabia, parts of Jordan and Syria towards western Iraq. Its basement consists of crystaline rocks, overlaid by sandstone, limestone and partly Cretaceous shales. Sands and eolian deposits cover most of the region. The Jordanian, Syrian and Iraqi areas are separated from the Mediterranean influence by the great Rift and mountains.

The *foothills and mountains* separate Mesopotamian Iraq from Turkey and Iran. They are younger and more complex than the Arabian shield. Formed by upward thrusts and folding in the Tertiary, they consist mainly of Cretaceous and Eocene limestone, pierced in the mountain crests by igneous rocks. The syn- and anticlines of their folding processes are visible in longitudinal lines on space photographs (*see Fig. 14*). Seas advancing and retreating caused wide areas of the Near East to be underlaid with layers of salt and gypsum; in the Jurassic the sea became shallower, bays were formed with deposition of organic deposits in north-west Iraq, including oil bearing formations. Drier climates in the Miocene caused the further deposition of marine minerals, bays became marshes, which were gradually drained by streams beginning to run down the slopes of the earlier formed mountains.

These water courses, streaming down the mountains, have created the third region, the *alluvial basin* which forms the centre and most important part of Mesopotamian Iraq. This work of rivers, continuous since the Tertiary, has filled the depression between the Kurdish and Zagros mountains in the east and the desert shield plateau in the west, with a deep layer of sediments, transported from the uplands into the plain with stones, gravel, sand and silt deposited in succession. Present Iraq is created largely by the river-system.

1

The truly remarkable story of the Mesopotamian alluvium will be treated in Part II.

Morphology of Mesopotamia

The most adequate description of the Mesopotamian land was given by the archaeologists of the Oriental Institute of Chicago in the volume on excavations in Iraqi Kurdistan (Braidwood, 1960). They called the wide arc of hills, mountains and the desert plateau an *'amphitheatre'* surrounding the river plain opening to the southern sea of the Persian gulf, with the Euphrates and Tigris cutting along their slope into the delta. A map (*Fig. 1*) allows to visualize the scene, which is the setting of so much historical turmoil.

As a result of geological structure, altitudes up to 3000 m above sea level gradually descend to 200 m enclosing the river basin in its middle and southern part, where it falls from 200 m to sea level in a gentle slope. There is only one moderate breach in this amphitheatrical arc in the north-western corner of the Levant and Syria. This was the pathway of invasions, conquering armies west- and eastwards, and a route for trade and travellers.

The present state of Iraq occupies 450 thousands of square kilometers of this vast area shared, as mentioned, by the three regions, but the desert part, occupying almost two thirds of the county should again be brought into focus.

Climate of Iraq

Obviously, with the arid desert-steppe looming so prominently, it is the climate which plays a decisive part. Rainfall and temperatures are the usual main ingredients of climate. Rainfall at two crucial stations, Mosul in the hills and Baghdad in the plain are demonstrated, together with their annual trend of temperatures and relative humidity in a graph (*Fig. 2*). A further illustration is given by a map of isohyets over the whole of Iraq together with a simplified distribution of the main soil groups (*Fig. 3*).

From this documentation the main facts of the climate emerge. Rain is strictly seasonal and regional: it is confined to late autumn and winter, although occasional freak storms may occur. The amount of rain is strictly regional and dwindles down from 1200 mm per year to less than 100 mm over most of the country. A similar distribution of rainfall governs the whole Near East from Israel, Jordan, Syria to Iran and further.

Temperatures vary enormously between summer and winter, as shown in the graphs from Mosul and Baghdad. During the summer they may soar to over 40°C and fall to freezing point in winter. A longitudinal section through the country from north to south is given in Table 1.

Relative humidity varies from 50–75 per cent in winter and falls below 30 per cent from May to October, indicating the hot and dry summer months. Strong winds occur in southern Iraq and may turn into dust storms, blowing mainly from north-west. Evaporation may reach up to 3000 mm in summer especially from irrigated crop-land. It affects also the levels of the many shallow inundation basins along the rivers and the marsh lands.

2

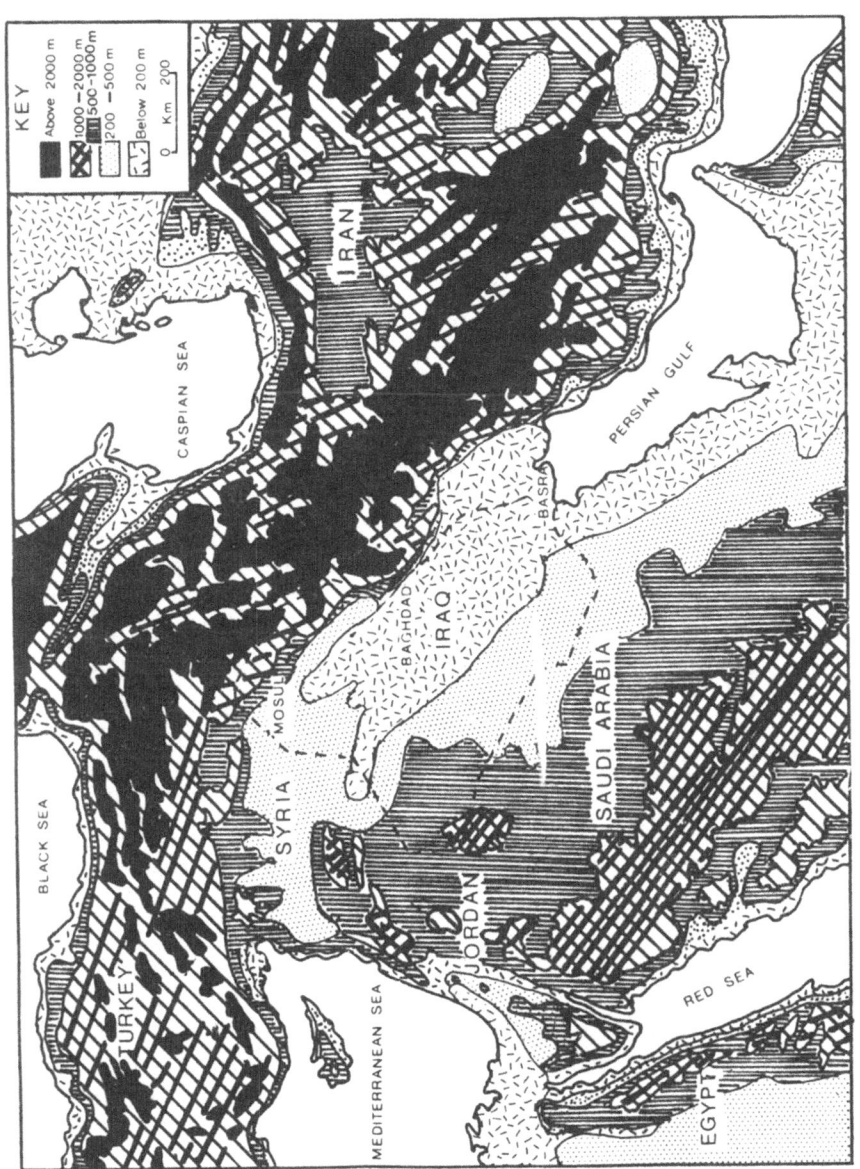

Fig. 1. Panorama of the Mesopotamian basin. The amphitheatre of mountains and foothills surrounds the alluvial plain sloping towards the Arabian Gulf. Note the lower passage through Syria famous in historical events. (From various sources.)

3

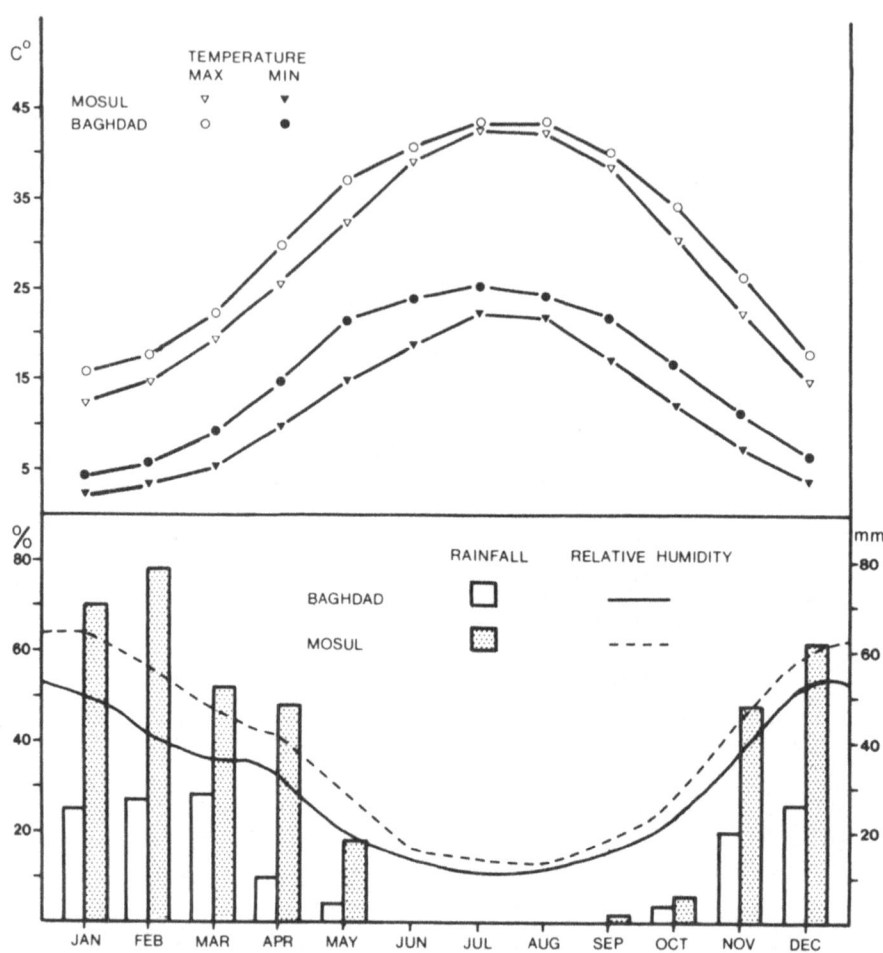

Fig. 2. Climate of Iraq. Temperatures, rainfall and relative humidity at Mosul in the hills and at Baghdad in the plains. Note the strong seasonal changes. (After Guest 1966 altered.)

Table 1. Temperature variation of Iraq.

Place	Temperatures in °C	January	August	Rain mm/year
Mosul	av. maximum	18.8	45.3	363
	av. minimum	12.1	12.7	
Baghdad	av. maximum	23.3	48.8	144
	av. minimum	7.7	18.3	
Basra	av. maximum	27.2	48.0	141
	av. minimum	4.4	20.0	

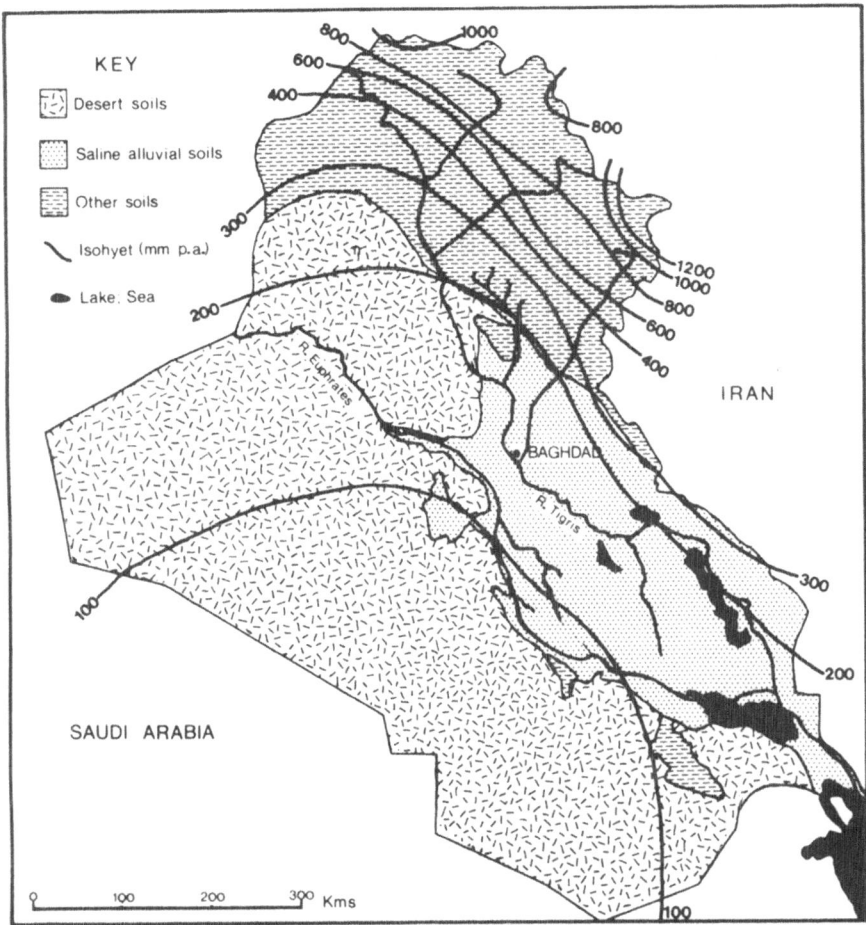

Fig. 3. Distribution of rainfall and soils. Isohyets range from 1200 to below 100 mm, reflecting the morphology. Note salinity of some soils. (Adapted from Guest, 1966.)

In conclusion the seasons govern the climate of Iraq decisively; rainfall is untimely in the cold months mainly, hot summers are practically rainless. There are in addition great fluctuations from year to year, from 50–450 mm at Baghdad. It is clear that those conditions affect severely the life cover of the country.

Soils

The composition and quality of soils are the result of geology, morphology and the climate. To this must be added the effect of irrigation in many countries relying on it. The map (*Fig. 3*) shows a simplified distribution of soils in Iraq, based on the detailed studies of Buringh and Kadry (1956) and Buringh (1960); these are quoted here after Guest (1966). There are 14 major soil groups, divided into more detailed subdivisions. For the purpose of

5

this book only four major regions are shown on the modified map (*see Fig. 3*), necessary to understand the effect of the land on the characteristics of the waters of Iraq. In the north brown mountain and hill soils prevail, showing the effect of higher precipitation and the eroded material of rocks, modified by the richer plant cover. These brown soils are fertile, as gardens and woodlands show around Kirkuk and Mosul and in the mountains. Desert soils occupy the greatest part of Iraq with differences according to location, from the Arabian desert to less harsh conditions northwards. There the desert soils change into dry and, further north, humid steppe soils. The fourth large category are the alluvial soils of the river basins. These are brought down from the mountains by river transport for millions of years. All categories are under the influence of the deeper strata with deposits of lime, gypsum, salt and other minerals.

These layers are dormant when complete aridity prevails but come to the surface by the action of water, even without the interference of irrigation practices. Thus occasional rains in the desert or dry steppe cause white salt patches to appear in wadis and depressions with occurrence of a scanty halophytic flora.

But according to authoritative sources it is watering by irrigation which is the main cause of the salinization of soils. Gabaly (1976) has assessed the extent of this effect in a wide range of arid countries which have to rely on irrigation, from Egypt to Turkestan in Soviet Russia. According to Gabaly 25 per cent of arable land in Iraq is afflicted by 'secondary', man-made, salination, largely through the insufficience of water in the hot season. Brinkmann (1967 and other studies) investigated the 'basis of irrigation economy' on the Syrian stretch of the Euphrates, where agricultural schemes have been set up in the alluvial basin from Raqqa to Abu Kemal. There salts would be sucked to the surface making some of the land unproductive unless large quantities of flood water is used effectively to wash away and drain the water. The surrounding countryside shows salt patches after some rains. The groundwater is strongly saline and affects the crops unless continuous watch and leaching are applied.

The groundwater has a powerful effect on the salt content of all waters in Iraq as will be discussed in Part II under water characteristics.

Irrigation has been practised in Mesopotamia for 6000 years and already an ancient text of 2000 B.C. mentions that 'the earth turned white'.

2 The response of the living world to the conditions

Vegetation and fauna

The present conditions for plants and animals have changed in the last 100 000 years (as described in chapter 3), especially under the influence of man.

Vegetation is a good indicator of all influences acting on a land and in turn vegetation is the primary producer, setting the conditions of existence for all consumers. The map on seasonal world climates and their vegetational zones by Troll and Pappen (1964) puts the area of Iraq as part of the broad sweep of desert and steppe region of north Africa across as far as Iran.

This broad alignment has been qualified by plant geographers in more natural details (Zohari, 1971). Two phyto-geographical provinces are discernible in the Mesopotamian area, the Irano–Turanian and the Saharo–Sindian provinces. The Irano–Turanian region occupies the larger part of Iraq; it is rich in plant species and their endemicity. The Saharo–Sindian province occupies Egypt, Sinai, Arabia and some parts of Palestine. It is poor in number of individual plants, monotonous in species composition and governed by greater aridity.

Both vegetation and its follower, the fauna, show clearly a transitional character between Africa and Asia with palaearctic influences.

In Mesopotamian Iraq Guest (1966) distinguishes 5 zones of vegetation, which follow strictly the climatic and surface structure of the land (*Fig. 4*). These are: (1) the desert and steppe, (2) the river-system and adjacent formations, (3) the alluvial land, (4) the Assyrian higher plains and Kurdish foothills, (5) the mountains.

The arid desert-steppe zone encloses the Euphrates valley and part of the Tigris. Two types of vegetation can occur in this vast area; one is seasonal and appears only after the winter rains in the form of annuals whose seeds are dormant through the hot rainless summer; they last only few weeks, but create during a time of exceptional rains the astonishing sight of a green desert, as I saw once when flying over the Nubian desert after a rare summer rain. The second category of plants consists of perennials, surviving either by deep roots, swollen underground parts or composed of xerophytes often of halophytic nature. Photographs by botanists, penetrating into the desert, show patches of such vegetation, especially in depressions where the roots reach the often saline groundwater, amidst bare salt crusts. It is remarkable that the nomadic desert dwellers, the Bedouin, can find fodder and plant fuel in this landscape. They have certainly contributed to the deterioration of vast stretches of the land. Absolute desert conditions occur only near Arabia, further north the steppe changes gradually from arid to humid conditions. Plants in this zone are richer but their life and growth is arrested twice, first by

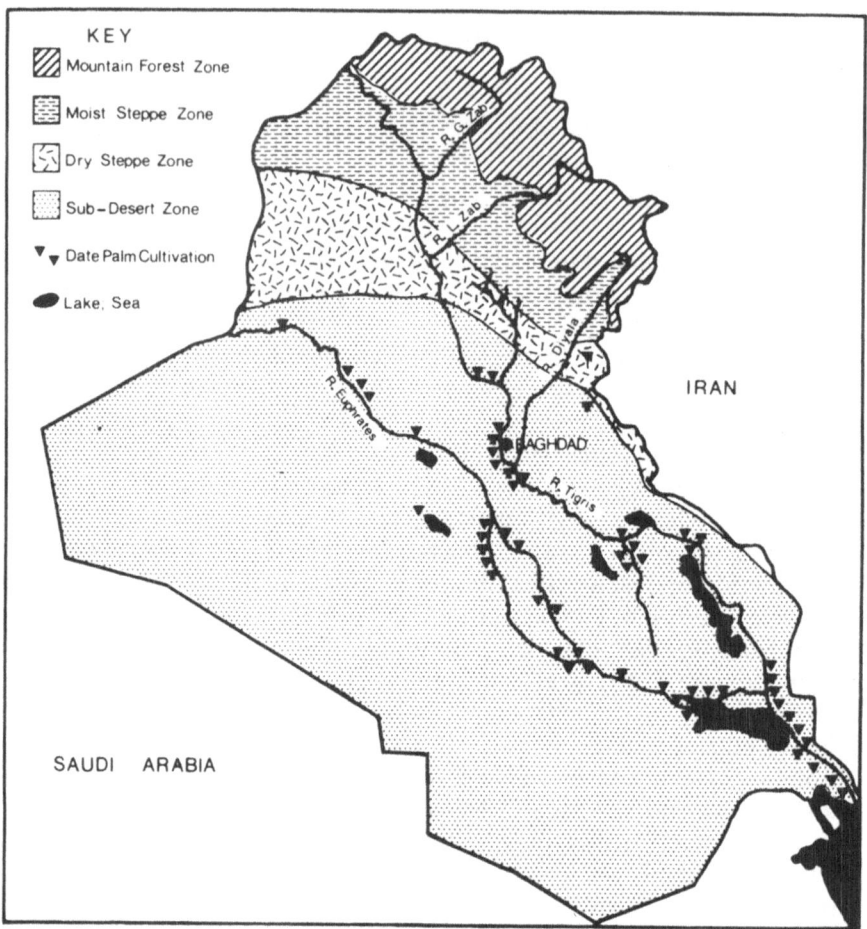

Fig. 4. Vegetation Zones and Distribution of Date-palms, reflecting dependence on rainfall and rivers. (Adapted from Guest, 1966.)

the cold of the winter and then by the heat of the summer. The physiological adaptations to these extreme conditions are dormancy, deep roots, reduction of leaves, tough cuticles and low bushy spreads. Large stretches are occupied by species of the Chenopodiaceae, often of 'marine' character or origin because of salinity.

The 'humid' steppe merges in the north with the Assyrian uplands and the Kurdish foothills. Though largely impoverished by thousands of years of man's activities of overgrazing, wood cutting and subsequent erosion, some vestiges of the native wild vegetation can be found and a much greater variety of species. Annuals live longer, perennials are not stunted, grasses can grow and amidst these are the wild varieties of food plants, whose seeds ancient man collected and selected by trial and error for his first attempts of farming; now these wild strains are collected by scientists for their genetic value.

In the hills perennials assume small tree size and this trend is accelerated in the mountains. Vegetation in the mountains varies with altitude, rain and slopes. Oak scrub becomes oak forest, much devastated, but now vigorously replanted (*Fig. 5*). Here the main tributaries of the Tigris show in their lower valleys riverine vegetation of *Salix, Platanus, Fraxinus*, grasses and shore plants like *Typha, Phragmites* and others, which reminded the Austrian botanist Handel-Mazetti (1914) of his native Danube 'Auen'.

Higher up in the hills the oak forests, greatly depleted, are now being restored; above 2000 m the peaks become treeless and only bushes of *Daphne, Euphorbia, Astragalus* and *Acantholimon* form thorn-cushions.

Fig. 5. Aqra road in the Kurdish foothills, showing depleted oak scrub. (Courtesy of Iraq Petroleum Co., London.)

The alluvial plain is cultivated intensively and 'hardly any vestige of natural vegetation remains' according to Guest. Along the river banks *Populus euphratica* and *Salix acmophylla* grow; near Mosul woods and forests of conifers and Tamarix exist and are now being protected against indiscriminate exploitation. At Baghdad irrigated gardens grow in profusion with groves of date-palms. But here and further south trees for timber are scarce as they were already in Babylonian times, attested by Herodotus and the bas-reliefs, in which log-rafts are depicted. The date palm grows along the rivers to about 35°N and forms veritable 'gallery forests' on the lower Shatt al Arab (*see Fig. 4*). Marshes have developed in the lower reaches of tributaries, when entering the plain and losing current force; a wide expanse of marshes has developed

9

since antiquity towards the sea in close dependence on sediment deposition and the advance of the delta; impressive space photographs show their extent and this outstanding phenomenon is discussed in chapter 5.

Whereas vegetation in the hills and mountains is rain-fed and the cultivated alluvium depends on irrigation, the marshes are created and maintained with considerable fluctuations by the rivers. Their extent and vegetation, as far as known, is treated in chapter 7.

In general, though the original vegetation has been reduced by overgrazing, wood cutting, replacement by crops and the salination effects of irrigation, vestiges remain in less populated regions. Botanically Mesopotamia and much of the 'Fertile Crescent' is rich in species; Boissier in his Flora Orientalis (1867–1888) enumerates 12 000 species. Guest (1966) mentions the great development of certain families and genera like *Centaurea*, *Allium*, *Salvia*, *Astragalus* and others, which appear in hundreds of species. Then there is the rich presence of wild grasses, ancestors of wheat, barley and other crop plants. The Mediterranean element is poorly represented and this is due to the extremes of climate and the intervening desert. Some species are apparently found in ruderal wasteland, possibly as legacy of trade, migrations and continuous passages of armies in the past.

Chapman (1960) in his book on 'Saltmarshes and Salt deserts of the World', has mentioned Mesopotamia. There are saline semi-desert areas in the Mahmudia region, near Babylon, with dried salt contents of 3.4 per cent, and pans with salt incrustations are conspicuous with the plant *Alhagi maurorum* dominant. Areas around Shithatha can have soils with sulphates and chlorides up to 7.5 per cent, usually barren, but blue-green algae may appear after rains; there are also stretches with *Suaeda* and other halophile plants, providing pasture for camels. East of Baghdad depressions bear patches of the grass *Aeluropus*, other areas are bare; salinity is high, but if below 1 per cent, annuals occur in mosaic pattern.

Crops and plants used by man

Judging from publications on Babylonian and Assyrian plant names, e.g. Bonavia 'The Flora of the Assyrian Monuments' (1894) and Thompson's 'A Dictionary of Assyrian Botany' (1949), ancient people paid attention to the plants around them. Food and medicine were derived from plants and at least in the medicinal knowledge there seems to be a continuity of the use of herbs. It would be of absorbing interest to compare the modern list of herbs used (Al Rawi and Chakravarthy, 1964) with those of the Assyrians.

Crops in the distant past included sesame, onions, a variety of fruit like apricots, peaches, apples, dates and figs; the vine was cultivated in the hills and exported. The most important crops were cereals, wheat and barley; their domestication was a long process from seed gathering of wild varieties recognized by about 7000 years B.C., to sowing, selection and gradual improvement of quality during more than one thousand years. Here the outstanding work of H. Helbaek of the University of Copenhagen must remain a milestone in palaeo-botany. Large parts of the Near East were recognized by him as the 'nucleus' of ancestors of our food plants; more about the fundamental process has been treated in the chapter on palaeo-ecology. The food debris of prehistoric man reveal some features of nature and incidentally show also the importance of inter-disciplinary cooperation. By the 4th century B.C. Herodotus was astonished by the crops of Mesopotamia as quoted in the Annexe of Eyewitness accounts at the end of the book.

As today, crops of the ancients were subjected to the conditions of climate; winter crops of wheat and barley can be sown in the Assyrian foothills in the north, summer crops in the plains are possible only by irrigation. The cultivable area of present Iraq has been estimated at about 20 per cent of the total area, most of this by the distribution of river water. This has lasted for 6000 years and has been referred to in several parts of this book as the outstanding feature of the country (see chapters 1 and 5). The instability of the river courses was sometimes destructive. According to archaeologists excavating Ur and Eridu, these thriving agricultural communities fell into ruin and abandonment when the Euphrates shifted its course to the east; Woolley's lament is quoted in the Eyewitness' Annexe. The land reverted to desert. Salination seems to be a result of irrigation practices as mentioned elsewhere (chapter 1). Another factor has contributed to the disappearance of the original plant cover and the deterioration of land in some parts. Jacobsen and Adams (1958) in an article on 'Salt and Silt in ancient Mesopotamia' have assessed the covering of the ancient land surface at up to 10 meters in the last thousands of years, burying the land and its settlements.

In Iraq the revenues from oil production, east of Kirkuk and in the Basra region, allowed government agencies to tackle the conservation of the natural plant resources and the better management of fringe lands. Two examples of such projects are documented in reports by Hunting Technical Services (1955 and 1958). The first deals with a 'pilot survey' of forests near Mosul, occupying an area of 541 square kilometres. Pine and oak predominate, growing on calcareous soils with a rainfall of only 380 mm per year. The exploitation was heavy, shifting cultivation, fires and grazing denuded the plant cover and increased erosion in the hilly countryside of the Greater Zab watershed. Sample plots were chosen, growth measurements made and the results and recommendations submitted to the Directorate of Iraq Forests. A rich photographic documentation includes riverine woods of *Platanus Fraxinus* and *Salix*.

The other, later, report deals with 'Conservation and Agriculture' in the Upper Diyala river, one of the most powerful tributaries of the Tigris. The lower parts are a rolling grassy plain with fertile alluvial soil, green after the winter rains, dried up to brown in summer. The country rises to 800 m near Sulaimaniya and higher upstream to the precipitous cliffs of the Qara Dagh range. Rainfall oscillates around 300 mm, temperatures range from 10 to 40°C. There is a short season for rain-fed crops and vegetation in winter and early spring; summer crops have to be watered by gravitation flow from streams in the hills, many temporary. Higher limestone ridges bear oak coppices amidst grasslands, most forests were degraded at the time of the survey. The boundary between humid and dry steppe lies at the 300 mm isohyet and the plant cover clearly reflects this present limit. This boundary shifted in the ancient past and forests and woodlands moved down as mentioned in chapter 3. Clearance of forests caused a secondary climax of grasslands. A large list of natural plant species contains wild varieties of wheat and barley, and other plants recognized by palaeo-botanists as 'progenitors' of domesticated crop plants. An array of photographs illustrates the report including a composite picture of the whole Diyala valley. Recommendations for the improvement of parts of the river valley are given. Incidentally these Kurdish hills have been the scene of the oldest archaeological sites found until 1960, as discussed in chapter 3. Quite recently a dissertation by D. C. P. Thalen on the 'Ecology and Utilization of the desert-scrub Rangelands of Iraq' has been published.

This concludes a survey of the vegetation of Iraq and some of the efforts made on conservation management. As mentioned over most of the country the plant cover has been denuded mainly by the action of man.

The first volume of the *Flora of Iraq* by E. Guest contains several hundred references and a general survey of the present vegetation.

Remarks on the land fauna

No comprehensive description of the fauna of Iraq has yet appeared, only papers on particular groups, and some are probably outdated. Therefore this chapter is fragmentary but allows some limited orientation on the character of *present* faunal conditions. As in the vegetation great changes have occurred, especially in the larger animals due to human pressure, a diminished plant cover and its shifts; some relevant facts are mentioned in the chapter on palaeo-ecology.

Earlier but still useful information has been collected and published by members of British forces during the 1914–1918 operations in this area. These appeared mostly in the 'Journal of the Bombay Natural History Society from 1920–1922'; it seems that this was the nearest zoological centre available at the time of war. Some of the identifications of species were later verified or changed in London.

Mammals

A report on the mammals of Mesopotamia by Cheesman in 1920–21 lists 36 species and contains mainly small burrowing animals of steppe character, gerbils and Meriones, Diplodillus; 8 bats, 2 hedgehogs, a porcupine and others. There are some larger carnivores, 2 Felidae, 2 Canidae, a hyaena, 1 mongoose, 1 marten; of ungulates 2 species of gazelles, noted then in large herds N.E. of Samarra, the wild sheep and the wild goat and boar. This list has been gradually enlarged by Iraqi zoologists. The lion should be mentioned specifically because it disappeared only in the 19th century. It was one of the most frequently portrayed animals in Assyrian-Babylonian bas-reliefs and according to the archaeologist Seton Lloyd (1955) was noted during early 19th century field work. In 1816 'three majestic lions' appeared on top of the ruins of Nimrud (p. 83); visitors to the British Residency at Baghdad in about 1830 were alarmed by the presence of leopard or lion cubs; in 1850 an archaeologist at Eridu, modern Warka was disturbed by a family of lions, which attacked his watchdogs (p. 161). Hence the tales of very old Marsh Arabs recorded by Thesiger (1964) of the presence of lions in the near marsh area are probably true. Thus the continuing distribution of the lion from north Africa across the Near East through Persia to its outpost in an isolated refuge in India is plausible. In Persia, asiatic great cats like the tiger and the cheetah may once have transgressed into Mesopotamia; but man's destruction of woodlands has sealed the fate of the tiger, the cheetah apparently survives as a few specimens in an Iranian national park according to Daresh Shuri (1978, Wildlife London, Sept. issue). A mixing of asiatic and north African elements is also attested by the dromedary and the Bactrian camel, the first apparently now used, the latter depicted on Assyrian bas-reliefs.

Birds

The early (1921–22) paper by Ticehurst, assisted by Buxton and Cheesman, is still informative, though now superseded by Allouse (1959). The first paper gives observations on the ecological distribution of birds in natural and man-made habitats like gardens, agricultural land, buildings and ruins, besides marshes, rivers, desert and foothills. In the context of this book only some major outlines are necessary. About 330 species of birds are listed and an interesting table arranges the species according to their stay in Iraq:

	Visitors			Migrants	
Residents	Winter	Summer	Passing	Stragglers	Not known
78	123	18	55	12	43

The ecologically best represented group are species affiliated with watery habitats; these form one third of the total 330 listed. Steppe–desert birds are also prominent. The outstanding feature is the great migration route from north-eastern Europe; 'countless myriads of water fowl' migrate to the waters of southern Iraq at present. The ostrich, 'long since exterminated in Mesopotamia', has in the past according to Prater (1920–21) extended its north African–Arabian range into Syria and Iraq; there is a representation of an ostrich in the excellent assembly of Assyrian–Babylonian bas-reliefs by Parrot (1961).

Reptiles and amphibia

Two early papers by Boulenger (1921–22) list 18 species of snakes, 20 species of geckos, agamids and varanids, lacertids and scinks, one species of *Amphisbaena*. Allouse has compiled a bibliography of 'Iraq and neighbouring countries' and its III contribution is on Reptilia and Amphibia (Bull. Nat. Hist. Res. Cent., Baghdad no. 6) and Khalaf (1959) wrote on 'Reptilia of Iraq and notes on Amphibia' in the same bulletin. I have not seen these publications.

Now numerous notes amplifying the knowledge of all vertebrate groups appear in publications of the Biological Research Center in Baghdad. The Amphibia should be of special interest in view of the richness of waters of the river system.

Fishes

Some lists of fishes have appeared in the recent past, but, as Banister in his contribution to this volume explains, only a strict revision can clear up many uncertainties of determination. Life in the waters of Iraq and some relevant features of adjacent areas are treated in the second part of this book.

Invertebrates

No comprehensive work on terrestrial invertebrates has appeared to my knowledge except a monograph on the Lepidoptera by Wiltshire (1957) published by the government. Otherwise there are scattered papers on Orthoptera, Hymenoptera and other groups. According to Uvarov, the well known locust specialist (1921–22), 66 species of Orthoptera live in the predominantly dry steppe-desert stretching from Arabia to Beluchistan. The fauna is said to be indigenous in this wide area. Too little is known about invertebrates to characterize it generally, though there may exist lists of species important for plant protection.

Most animal groups show the mingling of palaearctic African and Asian elements. Dr. Banister in his chapter on fishes has made pertinent remarks on the zoogeography of the fish fauna of Iraq waters, to which readers are referred. P. A. Buxton (1923) has given a delightful study of Animal Life in Deserts, which contains many references to Mesopotamia and is still a valuable source of information. Remarks on 'desert hydro-biology' can be found in part II of this book.

Insects as vectors of diseases

Through the kindness of Dr. Macan, I was able to read the book on 'Anopheles and Malaria in the Near East', which is a record of work carried out during the last war (1939–1945). As this important subject depends on the presence of water for breeding places of the aquatic larval stages, it will be treated in the chapter on hydrobiology.

The Handbook of Iraq (1944) contains two chapters, on vegetation and fauna, worth reading (pp. 181–205). Photographs of plant associations appear in the Handbook and newer ones in Guest's Flora, 1st volume (1966).

14

3 Palaeo-ecology of Mesopotamia

The past is the key to the present. In previous chapters the present state of land, climate, soils and the living world has been tentatively described. The question arises why and how has Mesopotamia become what it is today. The present Iraq lives by intensive irrigation agriculture and, recently, on oil resources. But early travellers have commented on the monotony and poverty of parts of the Mesopotamian plain and archaeologists like Woolley (1960) have contrasted the 'desolation' around ancient thriving sites like Ur (*see* Annexe). A deep reach into the past contributes to the answer.

Geologic aspects of the archaeology of Iraq

This was the title of a paper by Wright (1955), and there and in a later contribution to the Chicago volume (Braidwood *et al*., 1960), he reflects on climate and early man. The first paper is devoted to the history of the land since the Mesozoic. Three phases are distinguished: (1) The submergence of a wide area of the Near East by the sea from about 200 million years ago, extending into the Tertiary 60 million years ago. This phase produced the mineral layers of limestone, salt and gypsum; much later man profited from the flint, alabaster and building stone. (2) The second phase is the orogenic (Miocene) time, forming the mountains now separating Iran from Iraq. Volcanic eruptions in a wide area from Jordan to Syria provided man later with obsidian, used for tools and adornments. The mountains caught rain and rivers began to run down the slopes. (3) The last phase is mainly that of erosion, amplified by pluvial periods connected with the Ice Ages in the north. This Pleistocene phase with considerable climatic fluctuations caused shifts of vegetational belts with associated faunal appearances, the cutting of deep river valleys in the mountains and the transport of eroded material into the foothills and the plain. Gravel terraces, caves and alluvial fans were formed and became sites of early man's presence. With the extension of vegetation, caused by increased rains, water courses existed now reduced to dry or only occasionally watered wadis visible all over the Near East and often marked on maps (*Fig. 6*). Flint implements of early palaeolithic man attest to living conditions. In his contribution to the Chicago excavations Wright contrasts these early favourable conditions with the increased onset of aridity in the last 10 thousand years.

Prehistoric man in Mesopotamia

My present source of information is based mainly on the volume of the Chicago excavations (Braidwood *et al*., 1960) and other minor publications. Probably since then further results have been achieved which will fill some of the gaps. Butzer (1965) has reconstructed the physical conditions from Egypt

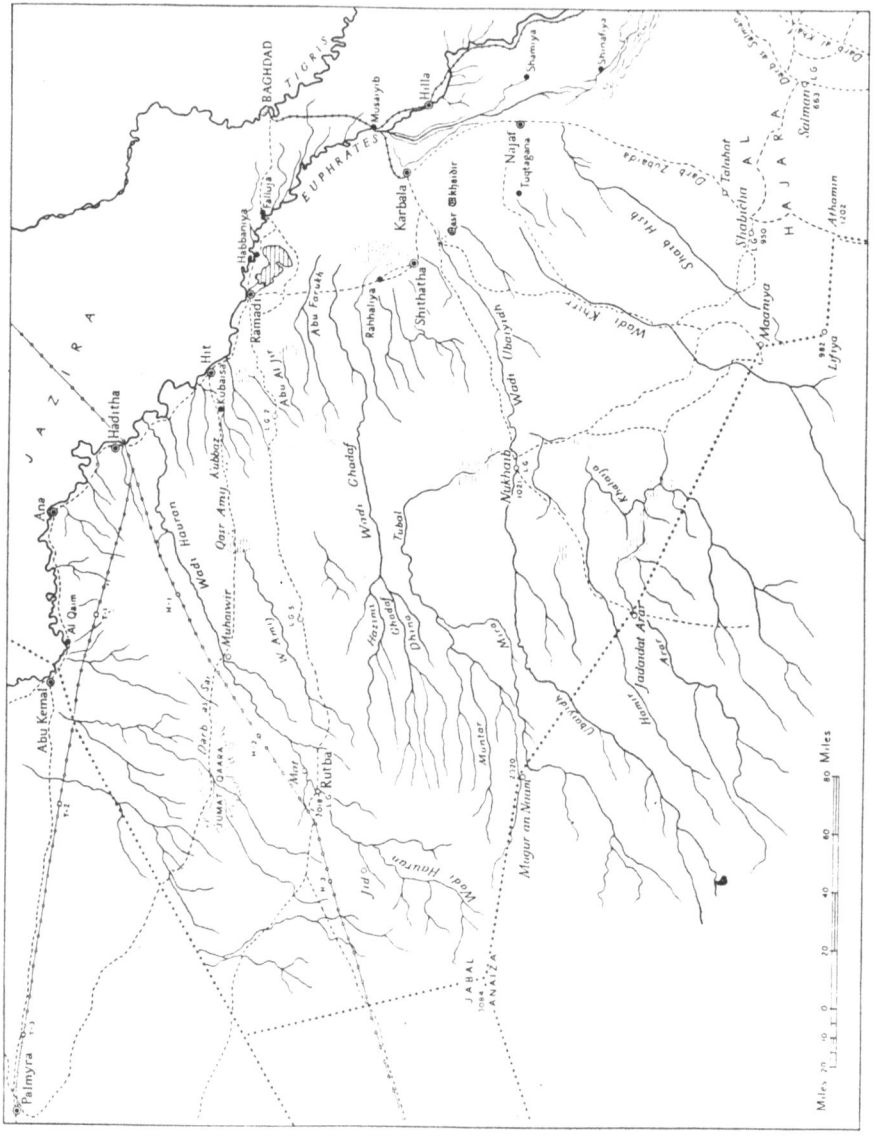

Fig. 6. Wadi System in the western desert of Iraq, ancient drainage channels towards the Euphrates. (From Handbook of Iraq, 1944.)

northwards with maps of presumed vegetation belts and appearance of fauna in the Upper Palaeolithic; he maintains that elephant, rhinoceros, hippopotamus and crocodile were present in the Levant and even into Arabia during the last, Würm, glaciation, 70 to 28 thousand years ago. Significantly all human sites were located in elevated grounds.

This is true for the finds of the Chicago work in Kurdistan (Iraq). The earliest traces of man in the present Iraq and elsewhere in the Near East have been revealed since 1928 by Dorothy Garrod, at Carmel and lake Kinneret* in Israel, Zarzi and Hazer Merd in present Iraq. Braidwood's team, guided by suggestions from the Antiquities Survey at Baghdad, excavated 16 sites in Iraqi Kurdistan, mainly east of Kirkuk and in the Chemchemal valley near Sulaimaniya. A detailed table of these sites has been shortened by me thus:

B.P.

75 000–100 000	Cave at Barda Balka; about 700 m altitude, present rainfall 320 mm.
50 000	Hazer Merd, 1000 m altitude, 400 mm rain.
	Shanidar, horizon D, cave in Aqra-Hatir valley, 800 m altitude, 600 mm rain.

B.C.

15 000–13 000	Zarzi and Palegawra sites, Sulaimaniya region, 1000 m altitude, 400 mm rain.
9000	Three sites, separated, 250–1000 m altitude, rain 320–600 mm.
7000	Jarmo, Chemchemal valley, 700 m altitude, 320 mm rain.
4500	Various sites, widely separated in Kurdish hills, up to 1000 m altitude, 270–600 mm rain.

I have marked the two main regions of prehistoric sites on the map of ancient settlements (*Fig. 7*).

The oldest site, to my knowledge, is the cave and adjacent stream gravel of Barda Balka. At present the archaeologists found grassy valleys, and hills covered by scrub oak. The stone implements were Acheulean (early mid-Palaeolithic), the faunal remains of their sites contained Indian elephant, rhinoceros of unknown species, wild ox *Bos primigenius*, wild sheep and goat, gazelles, an equid probably the onager and shells of the land snail *Helix salomonica*. In interpreting the faunal remains we must remember that they are what the archaeologists call a 'cultural filter', denoting the prey of man. Nevertheless, some conclusions can be drawn; elephant and rhino require

* O. Bar-Yosef has given a chapter on the prehistory of the Lake Kinneret area (C. Serruya 1978). A map (p. 449) contains the location of sites found around the lake from the lower Palaeolithic onwards to the Neolithic; the oldest site was dated at 640 000 ± 120 000 B.P. with Acheulean, very crude implements. Remains of man were assigned at later sites to *Homo galileus* and Neanderthaloids. This evidences to the wide spread of prehistoric bands of man in the Near East and to the results of a thorough examination.

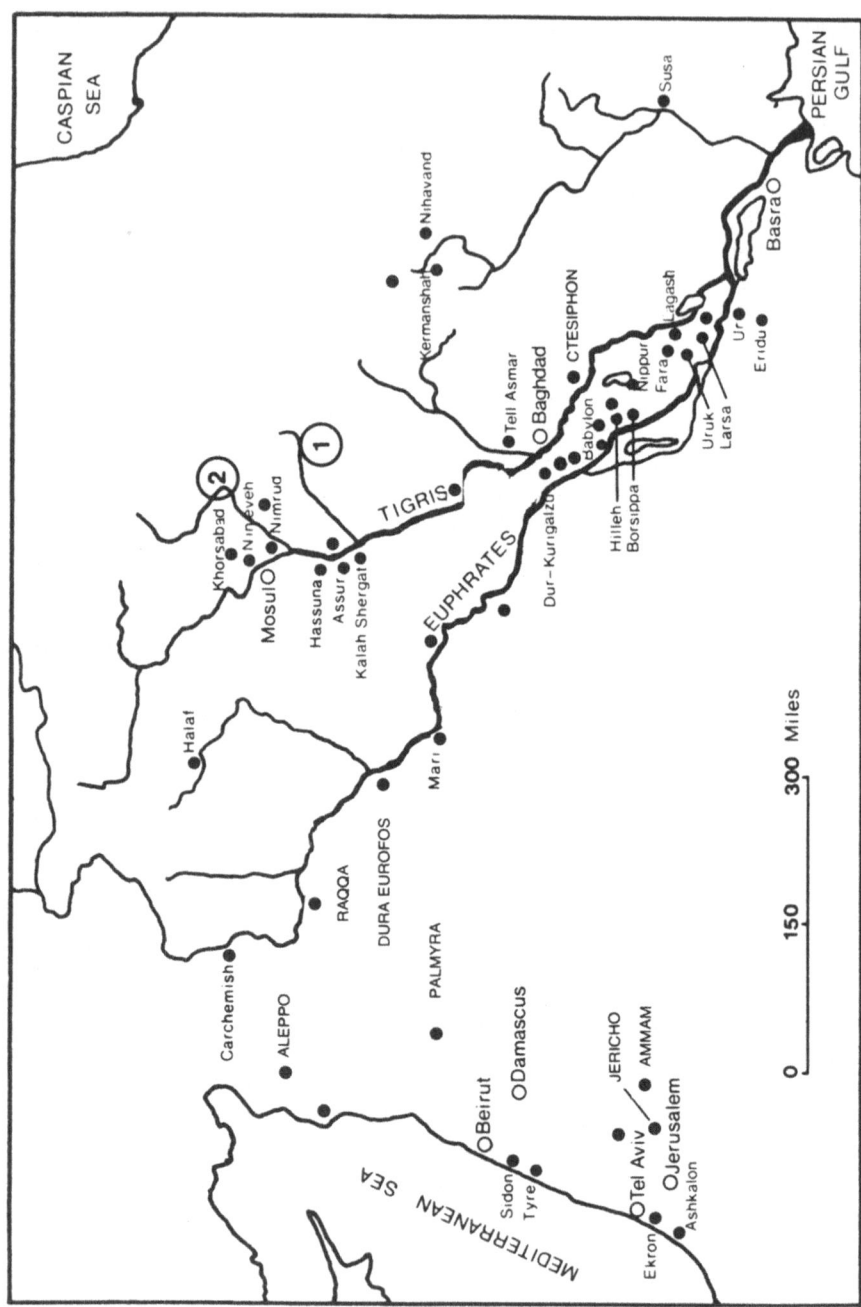

Fig. 7. Some larger Sites of Ancient cities. Note the confinement to rivers; nos. 1 and 2 are prehistoric.

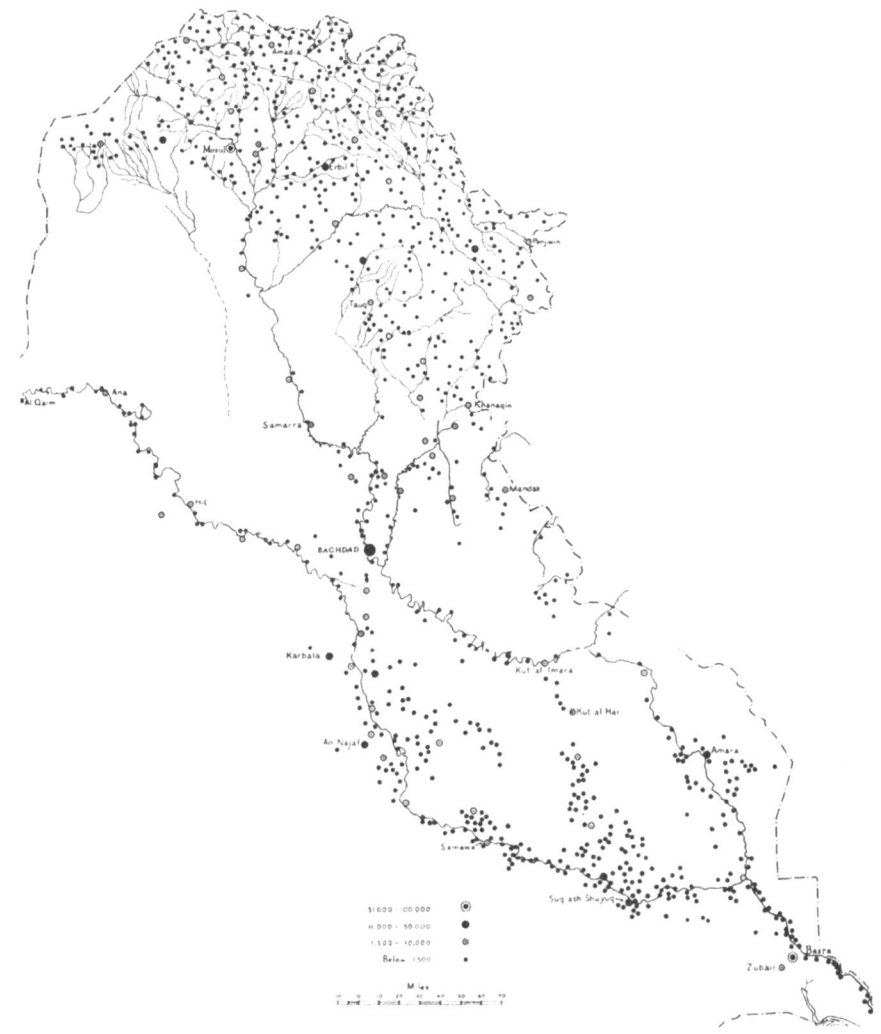

Fig. 7a. Modern Settlements showing the same dependence on the rivers and rainfall. (From Handbook of Iraq, 1944.)

large woodlands and ample grazing together with the wild ox and the gazelles. The shells occur increasingly as time goes by.

Hazer Merd and the lower horizons of the cave deposits brings us to Mousterien times, many thousand years later. In the Shanidar D horizon Solecki (1955 and 1957) found skeletons of Neanderthaloids with their characteristic stone implements. The faunal remains consisted of deer (*Cervus* and *Capreolus*), gazelles, the bear *Ursus arctos*, wild sheep and goat, wild pig and smaller mammals like gerbils, hares; birds; a tortoise and a number of snail shells appeared in the food remains.

The cave at Palegawra, about 15–13 thousand years B.C. yielded bones of

Equus hemionus the onager, a large bovid, *Bos* or *Bison*, wild sheep and goats, pig; the wolf *Canis lupus*, a fox, a lynx and small rodents were present, besides fish, toads of the genus *Bufo*, crabs *Potamon potamios* widespread in the Near East, clams *Unio tigridis* and numerous land snails. The inclusion of fresh-water animals from nearby streams seems significant. Plant remains consisted of pieces of large oak, poplar (*Populus euphratica*) and conifers.

The three sites dated 9000 B.C. had traces of seed collections and possibly 'incipient cultivation' according to the chronological tables of the Chicago excavations. Their primary objective was to find the beginnings of systematic cultivation of plants, that is agriculture. A re-examination of Jarmo promised success.

Jarmo, east of the present oil wells at Kirkuk, presents today the picture of a badly eroded silt plain at an elevation of 700–800 m altitude, but must have been in the past a more favourable habitat for a settlement of long occupation. The mound, containing the remains of this apparently thriving community, was 7 m high with a number of successive layers. The upper layers showed an outlay of houses, built of mud bricks of a special shape, used in a nearby village at present. Fire places were constructed from flat stones, polished by a nearby stream. Pottery became partly non-utilitarian and decorative, clay figurines of crude human and animal shapes indicated an advanced stage of life. Careful removal of layers revealed the sequence of stages; animal remains contained gazelles, wild sheep, goat and pig, the leopard (*Panthera pardus*), the wolf, a badger, marten (*Foina*), bear and small rodents. Freshwater crabs and clams were gathered and an astonishing amount of shells of *Helix salomonica*, a grassland species, now still found in small numbers. If the snail was a protein supplement to the diet, it showed an enormous change of habits; today no Moslem would eat it by tabu.

But the main search was for clues to the onset of seed gathering and cultivation. The palaeo-botanist H. Helbaek from the University of Copenhagen, examined the numerous plant debris. These contained the wild wheats resembling those occurring there now, *Triticum dicoccoides* and *Tr. agilopoides*; others are similar to *Tr. dicoccum* and *Tr. monococcum*. Barley grains found show a close relation to the wild *Hordeum spontaneum*; besides, the field pea, lentil, pistachio and acorns were present. Stone implements for rubbing and grinding, mortars and pestels, point to the use of vegetable food. In the beginning the seeds were probably gathered and much later sown.

H. Helbaek also took part in the searches of archaeologists of Michigan University. In a paper (1969) on 'plant collecting, dry farming and irrigation agriculture in prehistoric Deh Luran', these stages are examined. The site was an alluvial plain bordering an inundation depression of a stream. About 7000 B.C. two different groups of people lived there within a span of 700 years on the slopes of Iranian Kurdistan. The plant remains were meticulously examined by Helbaek in a truly remarkable piece of archaeological 'detective' work. A great number of plant seeds and other fragments could be determined as to species and it was assumed that some cereals were grown by inundation watering and channelling. It is through botanical expertise that there and in other sites the story of collecting to cultivation of some selected

20

species of wild wheat and barley in dry areas succeeded by primitive irrigation.

With this utilization of water resources people could move from rain-fed foothills into the alluvial plains. This step became important because it allowed the early cultivators to face more difficult climatic conditions. Two important chapters in the Chicago volume of excavation results are devoted to 'Environmental sequences in north-eastern Iraq' and to 'Climate and prehistoric man'. The chapters attempt to reconstruct the ancient environment with local shifts of vegetation in the last 10 000 years. No evidence exists according to the authors of these chapters that the climate has changed from present conditions of rainy, cool winter months to hot rainless summers. Thus people were forced to irrigation and its improvement. From about 6000 B.C. onwards man became the agent of environmental changes, which have lasted until now, wood cutting, over-grazing, and gradual salination of soils by continuous irrigation.

By the VI millennium sites appear near the great alluvial plain at Hassuna in the area of present Mosul and at Halaf near present Samarra. Skills of water management increase, demanding community compactness and some organization. At about the same time settlements appear in the south in the neighbourhood of the present great marshes. These precede the later developments of Sumer and have been named as the Al Ubaid period. Seton Lloyd (1960) said that there was an uninterrupted sequence from Al Ubaid cultures to Ur, Eridu and other sites. Dwellings of reeds were succeeded by remains of mud-brick temples, which attracted pilgrims and consolidated the importance of settlements also as market places. Now these are flat mounds only 5 m above the ground. In the same volume of the periodical 'Iraq' (vol. 22, 1960), Joan Oates asserts that connections existed between the Halaf sites near present Samarra and settlements in nearby present Iran; continuity of religious beliefs existed but a gradual amalgamation of semitic and other elements was in progress.

Buringh (1957) has tried to evoke the 'Living conditions in the lower Mesopotamian plain'. He regards the present environment in the alluvial plain as different in the past mainly through the combined action of the rivers and the aeolian sands from the desert. A layer of silt and sand up to 10 m deep, but usually 2–4 m, covers the ancient surface. He stresses the instability of the river regime as an important factor in the life of people.

Jacobsen (Iraq, vol. 22, 1960) traces 'The waters of Ur', the ancient river courses and canals, by the 'Ceramic Surface Survey' technique developed earlier in the Diyala region. Mounds (tells) stick out of the covering layers of silt and sand and are recognizable by sherds of pottery and other cultural debris. In this way not only human settlements but also the ancient water courses could be traced, remembering that in an arid desert climate people had to live near water, that is rivers and their derivatives. A sketch map in this paper (p. 174) shows a dense network of water channels from the site of ancient Nippur to Eridu, the Euphrates apparently having a more easterly course. A weir directed water towards Ubaid, Ur and Eridu. Clay tablets confirmed the existence of such installations.

The era of written evidence

Clay tablets present us with the invention of script and the voiceless past begins to speak to us. From the shadows of a largely mythological predynastic past, we enter in the literate phase of Sumer. Settlements became thriving communities, city-states based on trade and agriculture. Trade was by barter and large stores had to be kept, inventories organized. Thus from signs of goods a written version of the language came into being. The analogy with predynastic Egypt is striking. Thus the era of history has been reached; thousands of clay tablets have been deciphered, historical successions have been established.

The world of Sumer is not silent, archaeologists have unearthed treasures of evidence. We can now contemplate a human face from the Sumercan city of Erech (Uruk, present Warka) (*see Fig. 8*). Seton Lloyd, in his book on 'The Art of the ancient Near East' (1963), has remarked that marble in the Sumerian south was precious as there is no stone in that area. The face is actually only a 'mask' filled behind by bitumen, fitted to a wooden statue; it was painted and decorated by a golden wig and other decorations. The use of bitumen is interesting. It was found in several places of Mesopotamia and used for lowly processes such as binding bricks together, an interesting reflection on its importance today.

The Sumerians built 'ziggurats', tower temples reaching high above the ground; two explanations are offered for this type of building, security from floods and a religious tendency to worship towards the sky (*Fig. 9*).

Woolley, later Sir Leonard, excavating 'Ur of the Chaldaees' (1950), found a thick layer of alluvial silt between cultural deposits. He regarded this as evidence of 'the' flood, which reoccurs in the myths of ancient Babylon, Assur and the biblical legend. Now this event is seen as one of many disastrous floods which are a characteristic of the instability of the water regime of the river system. Even the main courses of the two great rivers changed in historical times. Ur is said to have declined when the Euphrates veered to the east; Nineveh was partly destroyed by a sudden flood; the southern marshes have changed their extent and the large Hor Hammar has appeared according to some sources only in the 6th century A.D. And then there is the problem of the delta end of the alluvial plain with disputes on its formation, advance and retreat. The space photograph (*see Fig. 25*) certainly suggests instability. This phenomenon of the Mesopotamian rivers will be referred to in the next chapter.

The era of written evidence contains the sequence of the spread of human settlements into the middle regions of Mesopotamia. It is not intended to give an account of the civilizations as they rose and fell; a multitude of books exist on Babylon and Assur. Here only some facts are discussed related to the role of man in influencing and managing the Mesopotamian environment.

Historic man in Mesopotamia

As would be expected, irrigation did preoccupy much of early historic man's thoughts. From predynastic time onwards, after 4500 B.C., many references

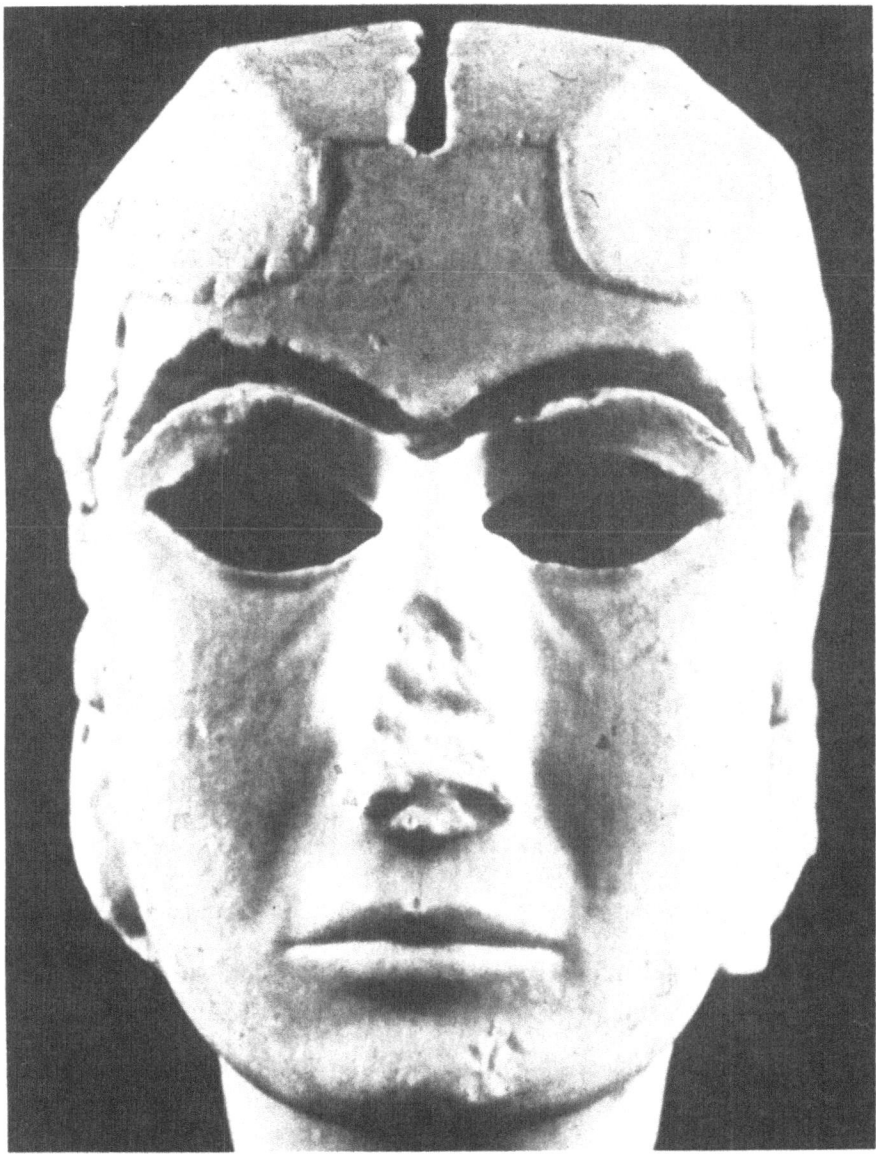

Fig. 8. Sumer, face mask of woman from Erech-Uruk, third millenium B.C. Scarcity of stone in the southern alluvium forced the sculptor to make only the front, the back was filled with bitumen. Once painted and decorated, according to Seton Lloyd (1961). At present in Iraq Museum of Antiquities. (Courtesy Iraq Petroleum Co., London.)

exist to bank and canal building, to drainage ditches, labour on these and legislation. Weirs were directing water to regions away from the rivers, elaborate networks of canals had to be made and maintained continuously. Assyrian kings extolled their achievements in water lifting devices; Herodotus in the 4th century B.C. contrasted water lifting irrigation with the basin irrigation of Egypt. The land lived by irrigation and after two thousand years

23

Fig. 9. Ziggurat at Ur, photographed in March 1954; now restored, Third millenium B.C. (Courtesy Iraq Petroleum Co., London.)

of this signs of salinization of soils began to appear; Buringh (1957) thinks that a Babylonian record from 1800 B.C. of 'fields turning white' refers to salinization.

Helbaek, examining the 'ecological effects of irrigation in ancient Mesopotamia' tells us of changes in the size of grains, seeds, brought down from the hills to the different climatic and watering conditions in the plains. Some wild species did not survive, others did. From Ur onwards to Nimrud, examined seed samples showed increase of grain size (e.g. wheat, barley, linseed). Wild wheat changed into emmer but improvements took a long time and there was only practice, no genetic selection. The salinization of soils exerted a considerable influence on the selective survival of cultivated species.

Trade was a very important feature of the life of ancient Mesopotamia, as seen in discovered port installations along the rivers. Harbour installations were prominent in the 'sea land' near the gulf, most probably identical with Sumer and the present marshland. In clay tablets control posts regulating sea trade are mentioned (Jacobsen 1960).

Internal river communication was common; there are numerous bas-reliefs at Babylon and in Assyrian ruins, depicting river scenes. *Figure 10* shows two types of vessels, a raft of timber (the kellek) and a coracle, the quffa, in water with whirlpools, waves, fishes and a crab feeding, observed with great accuracy. Herodotus was intrigued by the coracle-roundboat and described its

24

Fig. 10. Bas-relief from Babylon, about first millenium B.C. River scene with timber rafts ('keleks') and a coracle ('quffa') used for river transport; waves, eddies, fishes and river crab realistically depicted. (British Museum London; photograph by G. Fryer, F.R.S.)

construction from skins over a frame of wicker. A photograph (*Fig. 214*) in the Handbook of Iraq (1944) shows these quffas still in use.

Bas-reliefs show a wide array of animals mainly as prey of man; plants were apparently also of great interest to ancient man as catalogues of botanic names from monuments show, e.g. Bonnavia (1894), Meissner (1891), Thompson (1949).

There is an abundance of books on the ancient civilizations; because of the availability of stone in the middle and northern regions of Mesopotamia, Babylonia and Assyria have provided a wealth of bas-reliefs, inscriptions and tablets that allowed the reconstruction of ways of life, nature, religious beliefs, decipherment of poetry, political and trade documents. Thus e.g. Saggs could evoke 'The Greatness that was Babylon' in a book (1962) covering most aspects of that civilization: J. Hawkes (1963 and 1973), Parrot (1961), Butzer (1965), Woolley (1950) are further examples of splendid information on the Mesopotamian land of destiny. I have perused only some volumes of the two periodicals 'Sumer' and 'Iraq'; a thorough study of these sources could produce a much better palaeo-ecology than I was able to give in my present circumstances.

The tumult of history

I have called Mesopotamia a land of destiny and I regard it as necessary to add at least a brief chronology of the main historical events, which have swept through it and affected man profoundly since written records exist.

B.C.
4500–2170   Al Ubaid, Sumer, Lagash, Akkad-Lagad; predynastic and dynastic period. Rise and fall.
1792–600   Babylon, Assyria; empires and their downfall, invasions by Egyptians, Hittites, Hurrians.

25

| 519–150 | Persian conquest of Mesopotamia; Alexander conquers Persia, Seleucid successors. |
| 64 | Roman penetration begins; Pompeius takes over Syria. |

A.D.

105–164	Trajan annexes Arabia; trade and military outposts at Callinicum (near present Raqqa) and at Dura Eutropos, a previous Seleucid settlement, both on the upper Euphrates, connected with Palmyra by the strata Trajana, which the emperor Julian used.
224	Sassanid empire established with Ctesiphon on the Tigris as capital.
400–500	Mohammed and rise of Islam.
637–1055	The Arab caliphates; Arab conquest of Mesopotamia with capitals at Baghdad and Samarra. Crusades; Seljuk Turks emerge.
1258, 1393	Mongol invasions.
1534–1918	Ottoman rule and conquest of large empire.
1918	War 1914–1918 ends with dismemberment of Turkish possessions; western armies in Near East, mandatory rule, leading to delimitation of separate states.
1932	Iraq proclaimed independent kingdom.
1940–1945	Allied armies again in the Near East, safeguarding supply lines to Russia.
1958	Overthrow of monarchy, birth of republic.

These are only the major events in a much more complex sequence of wars, destruction of states, rise of others until the present.

Through this tumult people remained on their land, tilling it and continuing the succession, amalgamation and accumulation of races, languages and religions. The whole Near East with the present Iraq contains a mosaic of peoples, who once lived here and survived the various changes. As in Egypt there must be genetic strains of the ancient human elements here though overlaid by Arab influences. Chaldaeans persist with an ancient language of Aramaeic, which superseded Sumerian; Assyrians, Turks, Persians, Kurds mingle in northern Iraq with a number of old Christian sects; a number of different languages are still spoken in that area which was once the pathway of historical events.

The story of oil

The present wealth of Iraq, like that of adjacent Persia and the Gulf states, relies on oil which enables a.o. the rehabilitation land schemes now under way. This wealth was untouched until the first decades of this century when the Ottoman administration granted concessions to various British, French, German and other firms to explore and drill.

But knowledge of bitumen and oil dates back to the ancient world. The

Babylonian Epic of Gilgamesh, a tablet relating to the Deluge legend, contains apparently three words for oil products, bitumen, pitch and oil. Bitumen was used for mortar, pitch for caulking boats and pots, and oil for offerings and cosmetics. The ancient Persians coined the word nafta, the present Arabs used 'naft' for oil seepages; temples were built over such places in Elam and Persia for fire worship.

The Bible mentions it in its version of the flood; Herodotus and other ancient travellers saw the seepages and bitumen with astonishment in the last centuries B.C.; Marco Polo on his travels east through Mesopotamia and Persia in the 13th century A.D. mentioned the use of oil for burning and cosmetics. But only in the twentieth century did oil become a precious natural resource. This is a remarkable story of late social development.

Part II

Mesopotamian Waters
their
Regime and Hydrobiology

4 Near East waters as wider background

Mesopotamian Iraq is not isolated, it is part of a large area of the Near East. Leaving out the high plateau of Turkey, the waters of Israel, Jordan, Syria and Iran show some common features, like climate and soil salinity; bio-geographical affinities are also evident.

Four sites have been chosen, three of them west of Mesopotamian Iraq, lake Kinneret, the Dead Sea in Israel–Syria, the waters of the Azraq Oasis; and in the east, waters of Iran–Persia. Lake Kinneret and the Azraq Oasis lie in an almost straight line, sloping through the arid steppe-desert of Jordan towards the Euphrates basin. In the east the mountains of Kurdistan and Zagros separate the low alluvial basin of Iraq from the Persian plateau, but some of the Tigris tributaries come from western Iran. Here in Iran aridity and salinity of soils also prevails, culminating in the Great Salt Desert of central Persia. Lake Kinneret and the Azraq ponds and playas are the only waters from the Rift to Iraq.

Lake Kinneret

The lake is now in the political control of Israel though the eastern shores have been until recently Syrian. Names for the lake used, such as Sea of Galilee, lake Tiberias, evoke vivid historical memories and maps, descriptions and observations abound in past literature. Apparently 'Kinneret' is an old Jewish name.

As result of intensive all-round research over many years, a monographic study of 'Lake Kinneret' has appeared edited by C. Serruya (1978).

All aspects of the lake are treated by a number of co-authors: from geology, the age and origin of the rift valley, from East Africa through the Red Sea, Gulf of Aqaba, the Dead Sea and northwards. Water characteristics, hydrology, flora and fauna of land and water and man's presence are thoroughly discussed.

The book is one of the most competent, and possibly the best monograph of a lake.

Here only some few general and relevant facts are mentioned, as much as they fit into our aim. This is a true lake, not an inundation basin as are the standing waters formed by the Euphrates and Tigris. Essential data are: Area 167 km², maximum depth 43 metres, average depth 25 m, length 22 km, width 2 km; the shape is oval with little shore development. Its geological history is complex, with marine transgressions, underground infiltrations, a sequence of faulting, tectonic shearing and uplift since the Cretaceous. The glacial phase of the Quaternary is assumed to be the origin of the present water body, which is therefore regarded as no older than about 20 000 years. But there have been lakes before and sea transgressions, as testified by a relict fauna and the presence of man.

The Jordan river derives its freshwater from the snow melt and winter rains in the mountains of Lebanon; this water contrasts strongly with the lake water, which is much richer in Na and Cl, derived from numerous saline and often hot springs around and in the lake itself. To this must be added the effect of an evaporation of 1.80 m per year. Every year the river's spring floods cause a dilution of the lake water, with effects on the variation of its chemistry. Lake Kinneret is a warm-water lake with a range of 14–28.5°C; it displays a clear thermocline from May to November, and mixes in winter. The high mineral content is indicated by conductivity around 1000 μmho which varies with the season. Oxygen reaches super-saturation in the epilimnion due to algal blooms; depletion sets in the hypolimnion during the summer. Seiches both surface and internal have been recorded.

The living world of the lake is varied and specific. The phytoplankton is composed of some warm-water species and also forms of temperate and even cool-stenotherm character. A striking characteristic is the mass appearance of dinoflagellates, especially a species of *Peridinium*. The zooplankton has temperate-zone species like *Diaphanosoma brachyurum* and *Diaptomus gracilis*, the rest of the rather poor assembly is ubiquist. One would expect *Daphnia lumholtzi* here, a typical species of Middle–Mid East distribution, but it has disappeared (temporarily ?) around the middle 1950's.

The benthos is zoogeographically more interesting; we note here the presence of the Turbellarian *Dugesia salina*, at least 9 new species of Trematodes from fishes, the presence of Nemertinea of the genus *Prostoma*, Hirudinea of European–Western Asian affinities, some harpactoid copepods of marine-relict character, some new species of ostracods. Of special interest are Crustaceans of underground waters, relicts of an ancient fauna of marine transgressions: the blind prawn *Typhlocaris galilea*, *Cteniobathynella calmani*, *Monodella relicta* in company of subterranean molluscs and other forms.

The chironomids, 38 species or forms, are an important source of fish food; they are of palaearctic and African affinity. Eight species of molluscs live in and around Lake Kinneret, some with subterranean and marine past.

From our main point of interest in the zoogeographical boundary between Africa and Western Asia, fishes are the most conspicuous indicators. Of the 29 species known in Palestine, 19 species or subspecies are native to lake Kinneret. The fish fauna is transitional; African, oriental and palaearctic species mingle here, as an expression of the existence of the Miocene land bridge of the Levant between Africa and Asia. There are 9 African, 6 palaearctic, 2 oriental and two peri-mediterranean species.

Of the other vertebrate groups only few species are confined to the lake; they also represent the mixture of zoogeographic elements mentioned before. Man's presence has been attested around Lake Kinneret by numerous finds, ranging from lower Palaeolithic through all phases to the Neolithic; skeletons of a type described as Neanderthaloid and transitional to Homo sapiens were associated with Mousterien tools. At Ubaidiya, now on the Jordan river, but previously an old lake-site, the oldest human artifacts known in the Near East have been found and dated by various methods at 640 000 ± 120 000 B.P.

The Dead Sea

The Dead Sea is probably the most extraordinary waterbody in the world for geomorphological and ecological reasons, it lies about 400 metres below sea level, it is very deep, extraordinarily saline and, though not 'dead', extremely restricted in its life. The most astonishing phenomenon is the contrast between Lake Kinneret with its rich life, about 60 miles (96 kilometres) north along the rift valley, and the Dead Sea, which constitutes the extreme of sterility and of the salinity prevailing in the Near East.

The space photograph (*Fig. 11*), taken from about a height of 200 km, reveals some general features, such as the surrounding mountain walls, the sloping towards the east, where large wadi formations run down into the desert plateau of Jordania; these form the Azraq depression.

The Dead Sea lies towards the end of the great Eastern Rift, which extends from East Africa through the Red Sea, the Gulf of Aqaba into the valley, separating part of Israel and the whole of Jordan from the Mediterranean. The rift was created by subsidence (a 'graben') and orogenic movements in the Tertiary. The water level of the Dead Sea is deep below the sea level and its changes in the last 150 years have been recorded as between -391 in 1895 to -402 in 1977. The maximum depth is -750 below sea level, giving a slightly fluctuating maximum depth of the water of about 350 metres. As seen in the space photograph, the water basin is divided by a promontary into a (deeper) northern part and a (much shallower) southern basin. The total area is about 940 km². This great water body is endorheic, only a dried up wadi runs south as testimony of a former outlet. The river Jordan is seen entering the northern end but almost all water from it is now diverted. There are a number of salt springs and residual brines from former stages.

Limnological characteristics. A high evaporation rate and high air temperatures, combined with scanty rainfall govern the climate.* Water temperatures vary from 23 to 36°C, oxygen in the upper layers is 30 per cent of the normal, diminishing sharply with depth. Sediments contain, besides organic matter, iron-sulphide, aragonite, calcite and gypsum in shallow parts. The pH is 6.3 in the upper layers, 5.9 deeper down. Salinity ranges from 300 g/litre in surface waters to 332 g/litre in the deep. The ionic composition is:

Ca^{2+}	Mg^{2+}	Na^+	K^+	Cl^-	Br^-	SO_4^{2-}	HCO_3^-
16.86	40.55	39.15	7.26	212.40	5.12	0.47	0.22

g per litre

Mg. is the dominant cation, chloride dominates the anions, described as unique combination.

Biology. The alga *Dunaliella viridis* and possibly *D. parva* are the most

* However, there may be occasional rainstorms in winter; Yadin who excavated the rock fortress of Masada (1966, Weidenfeld & Nicholson, London) relates the occurrence of such violent rains sweeping over the Dead Sea area; dried up wadis were filled in few hours and flooded the archaeologists camp. Yadin tells of the ingenious way in which king Herod used such occasional water by leading it by gravity flow into enormous rock-hewn cistern. There are many examples of using desert rains, both ancient and present, e.g. in the Negev.

Fig. 11. The Dead Sea from space. Note the deep valley of the Rift, the entry of the Jordan, the descent of the eastern wall towards the Jordanian steppe-desert with wadis on the side. Quumran lies to the left of the Jordan entry, Masada to the left of the southern end, Ammam in the upper middle to the right of the rift. For the limnology of the Dead Sea see text. (Landsat.)

conspicuous organism, densities recorded at 4×10^4 cells/ml in the upper layers, few at 100 m. Sessile organisms *Aphanocapsa sp.*, *Phormidium sp.*, *Nostoc*, *Clostridium sp.* maybe brought in by some scanty Jordan inflow and small rivulets descending from surrounding hills. A cyst forming amoeba belongs probably to the *Dimastigamoebae*.

Bacteria form an important contingent of the living world of the Dead Sea, distributed according to chemical differentiation. To the halo-resistant group belong *Flavobacterium halmophilum*, *Chromobacterium maris-mortui*, *Pseudomonas halestorgus*; *Halobacterium maris-mortui* is halo-obligatory. The total number of bacteria decreases sharply vertically. The physiology of the principal organisms has been investigated.

34

There are no records of the presence of animals. A mosaic at Madaba east of the northern end of the Dead Sea shows 2 fishes, one dying the other trying to re-enter the Jordan. By sheer chance I found during a collecting trip the snail *Melanopsis praemorsa* and the large crab *Potamon potamios* in a streamlet near Qumran, at the north western corner of the Dead Sea. The Romans enlarged the little delta into a mosaic-covered basin before the entry into the totally hostile Dead Sea. This extraordinary occurrence is incorporated in a paper on the Azraq waters (M. Scates 1968). Finally a less scientific but significant observation; I saw people, intrepid enthusiasts, swimming in the Dead Sea without the possibility of drowning.

The above account was written from information supplied kindly by Dr. Colette Serruya, to whom I owe many thanks. The two main references are:

Neev, D. and K. O. Emery (1967), The Dead Sea. Bull. of the Geological Survey of Israel no. 41
Nissenbaum, A. (1972?), The Microbiology and Geochemistry of the Dead Sea. Microbial Ecology vol. 2: 139–161

Waters of the Azraq Oasis

To the east of the Rift valley the mountains of the Levant descend gradually towards the Jordanian steppe and the Arabian Shield desert. About 90 km east of Ammam, lie the only waters of substance until the Euphrates is reached. Yet this was a country of ancient sites, with Roman and Greek settlements, numerous lodges of early Arab rulers; wall-paintings bear delightful hunting scenes.

The oasis of Azraq lies in a depression of about 13 000 km² at 600 m altitude; but only 5 km² are occupied by permanent pools fed by springs, spouting from underneath the northern basalt shield into the limestone steppe. These spring pools, known to the Romans spill over into marshes surrounded by *Tamarix* bushes and reeds, flooded at times with islands of *Nitraria* (*Fig. 12*).

Winter rainfall, of about 150 mm, enlarges the marshes and fills numerous depressions over wide areas with shallow, temporary basins. Filled in December–January, these waters begin to dry out in April–May, when air temperatures rise from near zero to 40°C and evaporation reaches 10 mm per day.

There are thus two types of waters; the permanent pools and ditches and the temporary mud flat ('qa's') with distinct water characteristics and biology (*Fig. 13*). Electrical conductivity in the permanent spring pools keeps steady between 650 and 1200 μmhocm with water temperature around 25°C. Much of the marshes is covered by *Nitraria, Artemisia, Seidlitzia* in addition to reeds and *Tamarix*. The fauna of the permanent pools is poor in species, but locally rich in numbers of *Teodoxus marcrii, Hydrobia* sp., *Echinogammarus pungens*, besides five more species of molluscs and the crab *Potamon potamios*, 3 species of Harpacticids, of Near Eastern character, and ubiquist Cladocera and Copepods. A foraminiferan of the genus *Ammonia* astonished me, but was found by Löffler in Iran.

Fig. 12. Azraq Oasis, airview from the east; the dark area is flooded *Tamarix* marsh and Nitraria islands; temporary pools are visible in the background. The castle built here was once a Roman outpost.

Fig. 13. Permanent Pools at Araq, the only permanent waters, fed by springs, until the Euphrates is reached (except for some wells). (Figs. 12 and 13 copyright by E. Hosking, London.)

This fauna seems to be widespread in these regions. Only one fish species was found by me, *Aphanius dispar*, *Tilapia* was introduced. In the same semi-permanent ditches the shrimp *Athyaephyra desmoresti* was found.

Completely different is the fauna of the great number of temporary waters. These last only a short time, depending on size and depth up to 1 m deep and several km^2 in area. The mineral contents of these waters change rapidly through evaporation in high temperatures, up to 100 000 μmhocm were measured reflecting the nature of the soils in solution. Only a severely selected fauna can live in these conditions which I encountered also in rain pools in the Nubian desert around Khartoum (Rzóska 1961).

Euphyllopods dominate the scene, as one would expect; the genus *Triops*, *Branchinella spinosa*, a *Chirocephalus* and species of *Leptestheria* were numerous both at the time of my visit in April 1965 and found in April 1966 by Löffler and Bonomi. These Euphyllopods are ecologically significant but zoogeographically widespread. Gurney (1920–21) found this fauna around Baghdad and Amara in Iraq, as will be mentioned in chapter 7. The most interesting find was *Daphnia triquetra* (Sars) found previously in Russian Asia and in the Kara Kum desert; a more detailed discussion is contained in a paper based on my samples (Scates 1968). H. Löffler has given a detailed survey of the hydrobiology of the Azraq waters in a chapter of the report on the 1966 Expedition (Morton Boyd 1967). The main aim of this and previous visits was to lay down foundations for a proposed national reserve under the patronage of King Hussein. There are now only vestiges left of a former fauna of larger mammals; e.g. *Oryx* the cheetah and ostrich, to name some indigenous species, have disappeared; but a very rich avi-fauna lives there especially during the months of migrations. Reptiles are well represented, the mudfish *Clarias* has been found apparently, besides the marsh-frog *Rana ridibunda*.

The Jordanian Desert has been described by G. Mountfort with a wealth of superb photographs (1965).

The waters of Iran

Most of Iran consists of a high plateau from about 1000 m above sea level with mountain ranges of 3000 m and depressions into which endorheic short rivers discharge and where standing water bodies lie, of different degree of permanence and area. Rainfall, as all over the Near East, is seasonal in winter, followed by a dry, hot summer. The response of the vegetation is a steppe-desert association. Thus on maps of climate and vegetation the areas of Jordan, Syria and Iraq are similar.

There are also differences in Iran, foremost the absence of large central rivers, present in Iraq, further the presence of about 70 lakes formed by tectonic events, therefore quite different from the inundation basins of river origin in Iraq. Their size varies from about 6000 km^2 of Lake Urmia to small bodies of 1 km^2.

This section is based on the extensive work and the publications of H. Löffler (1956, 1959, 1961), who has brought together a rich bibliography on the geology, morphology and hydrography of the area. As in Iraq, sea transgressions were followed by the uplift of mountains; soils were permeated by salt and gypsum deposits and covered by later deposits of sands, gravel and other material. In this book we have to concentrate on the hydrobiology of Iran, which was Löffler's interest.

He has arranged the waters studied into three groups: (a) terminal lakes of endorheic water course, saline, shallow; (b) oligohaline lakes fed by springs and freshwater streams; both categories in mountainous regions; (c) lowland basins partly freshwater but in some cases rising to polyhaline concentration especially near the Gulf.

Of the many results, mainly concerned with the invertebrate water fauna, some examples may be quoted. Lake Urmia, in the north-western corner of Iran, is the deepest and probably oldest lake in Iran, possibly a relict of sea transgressions; at the time of Löffler's visit its water had a concentration of 366 g of salt per litre. The fauna was extremely poor with *Artemia salina* in great numbers and phytoplankton of 'small green masses' and some flagellates.

The most intensively investigated lake, during the expeditions (1959), was lake Niriz in the south-western region. Its area varies between 1200 and 1810 km² with a variable depth of around 1 m. It is fed by two streams descending from surrounding mountains (2000 m a.s.l.) to the basin at 1000 m above sea level. The mountains receive 1000 mm of rain, the lake only 400–500 mm; evaporation during the hot summer exceeds the supply from the affluents, hence the strong level fluctuations. The lake is constricted by a promontory causing considerable biological differences; the northern part is under the influence of freshwater inflow from the two streams and this water spreads over the denser saline waters, like the Jordan water in Lake Kinneret, but there in a much lesser degree. The salinity gradient is predominantly horizontal along the axis and amounts to 15–150% with halophile species largely confined to the southern basin and the general biological composition poorer. Löffler names *Mastogloia braunii* and a new species of *Nitzschia* amongst other saline forms in the phytoplankton, *Artemia salina* and halophile *Harpacticids* in the southern basin but not present in the northern part. The crab *Potamobius potamobius* and *Caridina fossarum* have been found in the area; for the genus *Caridina* this is the first site in Iran. The bottom fauna is poor with *Ochtabius*, *Ceratopogonid* and *Chironomid* larvae; in the littoral live 13 species of mollusks, and some *Hydrobia* forms with marine character. A rich assembly of Coleoptera populates the salt-encrusted shores, especially of the genera *Bledius* and *Dyschirius*, a Foraminiferan named as *Streblus beccarii* may be the same species as that found at Azraq. Twenty species of Rotifers and a number of mainly ubiquist Cyclopids are present.

Of great interest is the observation that Cyprinodont fishes, including *Aphanius sophiae* (Heckell), simply swarm in the delta waters of the two northern streams but die on the perimeter of the freshwater influx. A rich avifauna composed of European genera mixes with bee-eaters, flamingo flocks and other warm water forms.

In general the Crustacean fauna of the Iranian waters consists of European-Palaearctic ubiquists and some species of more significant zoogeographic affinities like *Diaptomus spinosus*, the *Cyclops strenuus* groups and with ecologically characteristic species like *Diaptomus salinus* and *Metacyclops minutus*; more detailed lists are found in the published papers. So far no relict forms of the former sea transgressions have been found which should exist in caves, springs and other underground waters.

An overall character of Iranian waters is the influence the saline soils have on the chemistry of the waters; Löffler declares e.g. the Niriz lake as a 'NaCl-MgCl$_2$-Na$_2$SO$_4$' combination and this description may apply to other water bodies; some have a prominent Natron character. Much of central Iran is occupied by the Great Salt Desert and this important feature explains the characteristics of the hydrobiological constitution.

Löffler drew attention to the impact of irrigation practices on the quality of natural freshwaters, during a MAB/UNESCO meeting on the 'impact of land runoff' (in press 1979). During the time when Persepolis was the capital of the Achaemenid empire, 2200 years ago, the streams supplying Lake Niriz and the lake itself were extensively used, until increasing salinization interfered with agriculture.

Waters of the countries surrounding Mesopotamian Iraq have been discussed and added to the background of environmental conditions. This panorama of arid lands and scarcity of freshwater is drastically cut by the two great Mesopotamian rivers forming a long enclave of water abundance and intensity of human life. The rivers dominate the scene of Iraq.

5 Rivers of Mesopotamian Iraq as dominating factors

No other area in the whole Near East has such powerful water supply as Iraq. Mostly small rivers exist in the Levant, directed mainly into the Mediterranean, in Turkey and Iran rivers of substance flow mainly into the Black Sea and the Caspian. The abundant supply of snow melt and winter rains in the mountains around Iraq is channelled by geomorphological features towards the south and forced to flow into the Persian Gulf. This unique arrangement dominates not only the geography but also the history of Mesopotamian Iraq. The Euphrates and the Tigris with their tributaries are life arteries.

Origin, length and slopes

Both the Euphrates and the Tigris arise in the Turkish High Plateau near Erzerum at an altitude of over 2000 m above sea level. Numerous streams join, to form into the two main rivers. The tributaries descend mainly from the east, the Kurdish and Zagros mountains, between Iran and Iraq. Most water comes from sources outside Mesopotamian Iraq.

The Euphrates is about 2600 km long to its (previous) junction with the Tigris at Qurna. After making an arc through Syria the river breaks through the western desert shield in a gorge and enters the alluvial low basin. Riverine depressions formed by river action extend in Syria between Raqqa and Abu Kemal, and are used for irrigation schemes. This region was prominent already in antiquity as a pathway to the east, as Roman military outposts show. In the plain the river loses its force and begins to meander in its own alluvial bed, a space photograph testifies (*Fig. 14*).

In Iraq the Euphrates falls from about 250 m a.s.l. to 4 m at Qurna over a distance of 1400 km or 0.18 m per 1000 m. I have drawn an approximate profile from some data given mainly by Guest (1966) (*Fig. 15*). Buringh (1957) has previously drawn profiles of the Euphrates and Tigris, based on

Euphrates			Tigris		
	Miles			Miles	
Stretch	Direct line	By river		Direct line	By river
Abu Kemal–Ana	59	100	Pesh Kabur–Mosul	68	125
Ana–Haditha	35	60	Mosul–Qala Sharget	60	73
Haditha–Hit	42	70	Qala Sherg.–Tikrit	64	78
Hit–Falluja	58	82	Tikrit–Baghdad	100	143
Falluja–Hindiya	53	80	Baghdad–Kut Imara	103	213
Hindiya–Samawa	115	150	Kut Imara–Amara	90	126
Samawa–Nasyriya	60	85	Amara–Qurna	60	90
Nasyriya–Qurna	70	85			

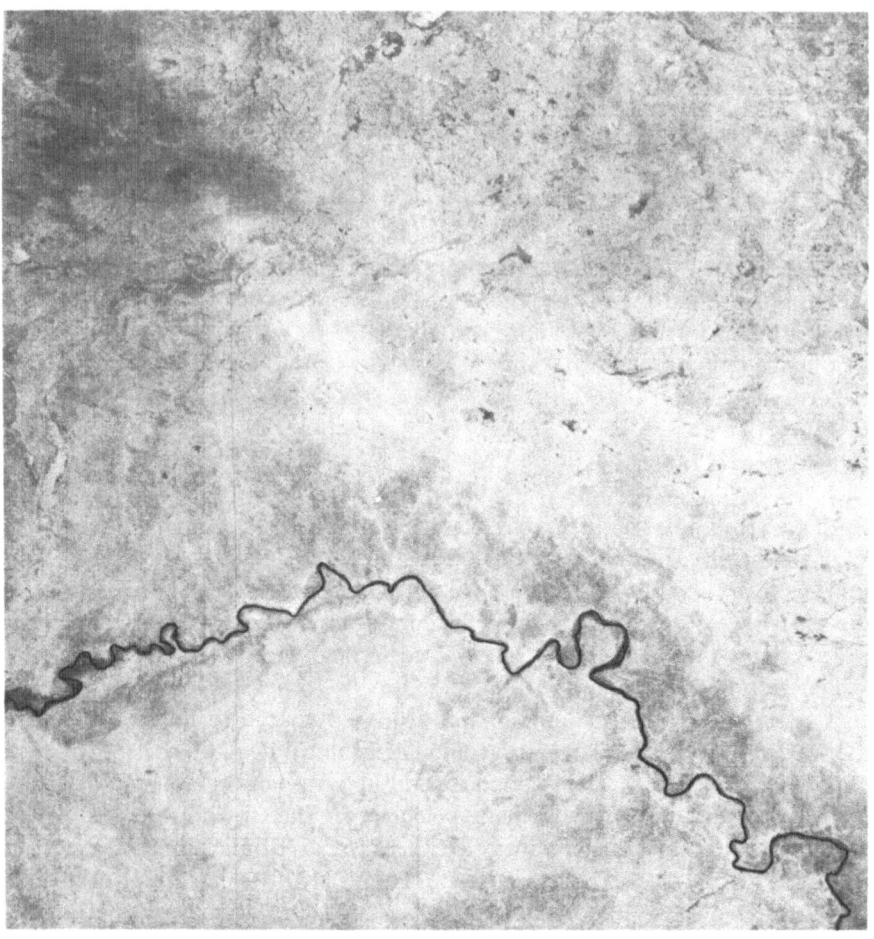

Fig. 14. The Euphrates on entry into the Iraq plains, region of Abu Kemal-Ana. Note the meandering. Along the river narrow alluvial beds are cultivated by irrigation. Space photo by Landsat.

falls in consecutive sectors. The first river falls 0.30 m/km between Ana and Hit, only 0.10 m/km and even lower in downstream stretches. The Tigris has a slope of 0.56 m/km at Mosul, but below Baghdad only 0.07 m/km; such low slopes cause meandering and other deviations from a straight course.

The Handbook of *'Iraq and the Persian Gulf'* (1944) contains observations on distances of river stretches both in straight line and by river in Iraq (*see table on p. 41*).

The balance is

	Direct line		River	
	miles	km	miles	km
Euphrates	492	787	712	1139
Tigris	544	870	848	1356

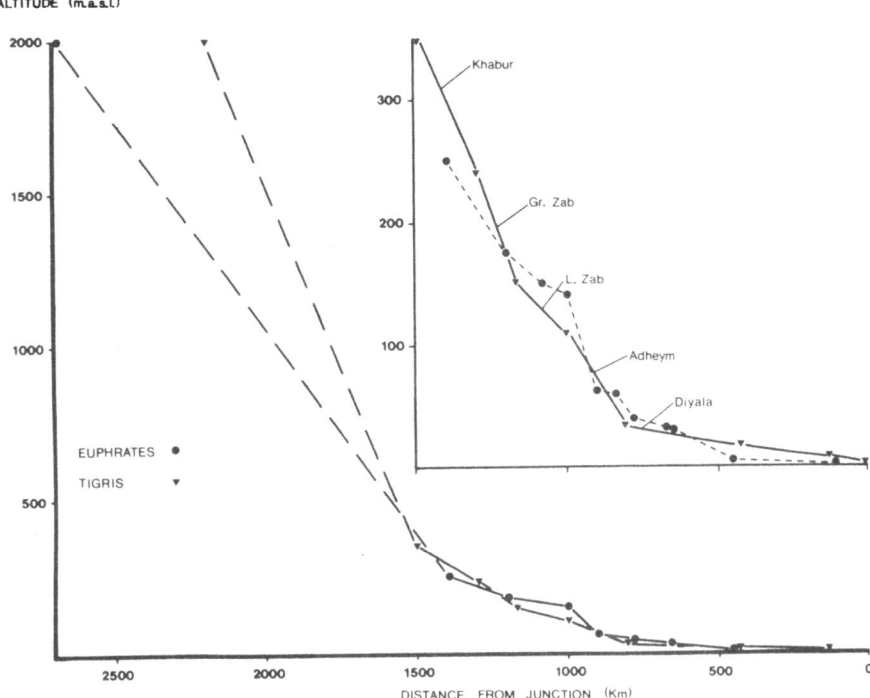

ALTITUDE (m.a.s.l.)

DISTANCE FROM JUNCTION (Km)

Fig. 15. Slopes of Euphrates and Tigris, drawn tentatively from existing sources; slopes govern currents and the capacity of carrying sediments. Note the slight descents towards the junction of the two rivers and the Shatt al Arab. (Drawn from data in Handbook of Iraq, 1944.)

Both river courses are prolonged by meandering and this in turn shows the loss of force of the water with great effects on the river vallies. Further we can understand why river communication was laborious and difficult before the steam age and why land communication became preferable.

With the loss of force due to its flat slope, the lack of tributaries, the Euphrates begins to dissipate into small branches, which in turn form marshy depressions. e.g. at Hilla and on a much larger scale east of Nasiriya, where a series of basins form the great southern marshes. These involve also the lowest course of the Tigris.

Maps of the western desert enclosing the Euphrates show a large array of wadis, at present dry or only exceptionally carrying rain-storm water; these extinct water-courses were alive during the pluvial times following northern (European) glaciations; stone implements of prehistoric man have been found in a large area of the desert plateau (*see Fig. 6*).

Some sources quote a drainage area of the Euphrates; this seems to be purely theoretical as over most of its course the river does not receive any supply; this is similar to the last 2000 km of the Nile, where the river is enclosed on both sides by sheer desert.

The Tigris is nearly 2000 km long, of which 1360 km run through Iraq. Its

43

course is dictated by the hills and mountains of the Zagros chain and of Kurdistan. As expected the supply of water from the snow melt and rain catching mountains is big. From north to south five tributaries drain the mountains: the Khabur, Greater and Lesser Zab, the Adheym and the Diyala. All these rivers, with the initial force provided by steep slopes, carry their erosion products into the plain, where they join the Tigris (*Fig. 16*). Thus the drainage area of about 166 000 km^2 is real and this constitutes one of the differences between the two Mesopotamian rivers. All these tributaries lose their force in the plains, form deltas, depressions and marshes. In spite of the powerful, though strongly seasonal supply, the Tigris, like the Euphrates, loses its impetus downstream and in both rivers the volume of water diminishes downstream. A flat slope, a strong consumption of water by irrigation and dissipation in its last stretches make the rivers weak.

The width and depth of both main rivers depend on the seasonal discharges and the morphology of the river-bed. Depths of 17 m as at Hammaniya south of Baghdad are exceptional and on the whole both rivers are shallow, from 2 to 5 m deep. The width varies even more under the influence of the spring floods as testified by large-scale inundations and flood disasters in the past, at Babylon and Nineveh, and some recent ones. At Baghdad sudden rises of up to 6 m have been recorded recently. Such flood rises occur also in the Tigris tributaries especially in the Lesser Zab and dams have been built in recent years on the Diyala, and on the Lesser Zab in the mountains (*see Fig. 16*). The regime of the river system will be treated in the following pages.

An outstanding characteristic of both Euphrates and Tigris is the instability of their courses. There are ample records of changes in historical times, as at Ur and Eridu for example, and some ancient river courses can be traced even now; some have been restored and are used as supply and navigation channels e.g. the Gharaf. A widespread old and modern canal system transects the alluvial plain as sign of the necessary transport of irrigation water; the rivers in Mesopotamia are arteries.

The most important tributary of the Tigris seems to be the Diyala, descending from the Kurdish mountains and entering into the main river below Baghdad. In connection with recent agricultural projects, a detailed ecological study was undertaken (Hunting Technical Services 1958). The river descends from altitudes over 2000 m high and is fed by numerous small water courses from Iran and some others further down until its entry into the plain at 35 m a.s.l. A reservoir at Derbendi-Khan has been built in recent years, to store the waters, south of Sulaimaniya about 800 m a.s.l. The upper regions of the river valley receive adequate amounts of rain in winter–spring but, except for sporadic rainstorms, nothing in summer; many water courses dry out.

As in most parts of Iraq, there is a short winter-rain fed growth season for vegetation and crops; from early summer until autumn the vegetation becomes dry and brown. Hence wherever the terrain allows irrigation has been introduced. This is possible on the terraces of the river valley, up to 100 m above the present river bed, and in alluvial silt beds formed in the deltas of some of the tributaries of the Diyala. Three geological formations, Eocene limestone, Miocene red-brown clays and sandstone, and Pliocene conglomerates, clays and silts mingling with more recent Quaternary gravels and silts, impart their different solids and solutes to the main channel and thus

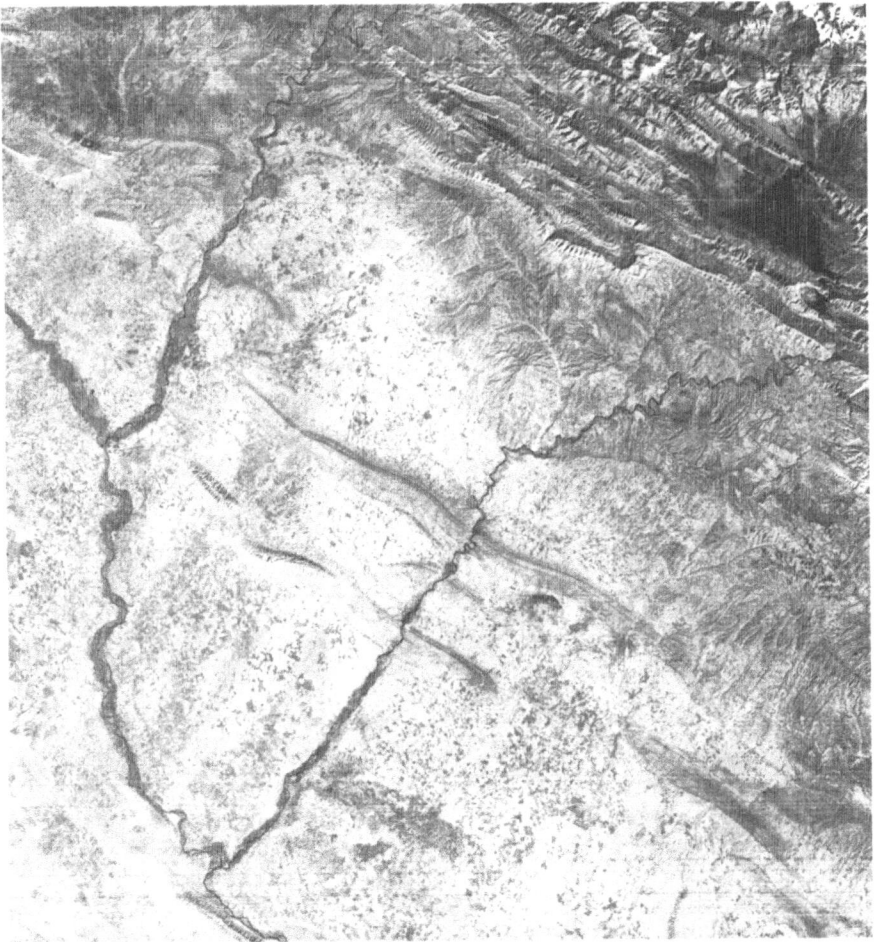

Fig. 16. Tributaries of the Tigris descending from Kurdistan mountains; the Greater and Lesser Zab have to traverse rocky folds of mountain ridges formed by orogenic processes in the Tertiary. The new reservoir on the Lesser Zab is visible. (Space photograph by Landsat.)

to the Tigris. A considerable load of sediments, almost 12 000 m³, are carried into the plain.

This example of a tributary has been quoted; the same conditions exist in the other main tributaries. Some parts of the river valleys are now despoiled, the forests have been denuded. Past conditions in these mountains and hills were more favourable, as evident from prehistoric sites (see chapter 3).

Standing waters

There are no natural lakes, in the limnological definition, in Iraq, but two reservoirs in the Kurdish mountains at Dokan and Derbendi-Khan show features of lake conditions. Dokan has an area of 270 km^2, a volume of 6.8 km^3 but with great fluctuations of levels, thermal stratification and considerable transparency (Al-Hamed 1976); a space photograph shows its location and shape (*see Fig. 16*). According to Sahaf (1975) the Derbendi-Khan reservoir has an area of 121 km^2; both Dokan and Derbendi-Khan are about 85 m deep.

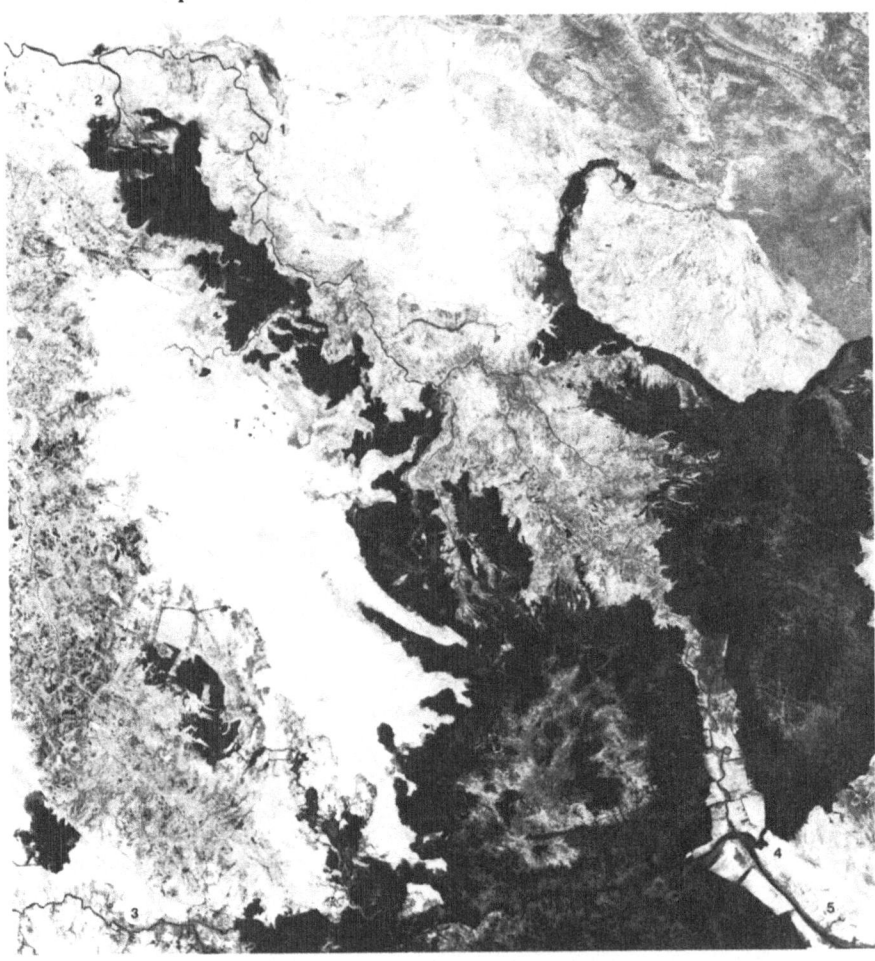

Fig. 17. Space View of part of the Southern Marshes, region of Kut al Imara down to the old junction of Euphrates and Tigris at Qurna; the Tigris is visible in the upper left corner of the picture together with a side arm, the Gharaf, possibly an ancient channel; the Euphrates is visible running along the lower margin of the picture. The Qurna junction, now less functional, is shown on the lower right side. Note the weakness of the Tigris, hemmed and dissipated by the marshes (dark areas); note the numerous drainage channels between the two dark marshes. (Landsat.) 1. Tigris below Kut al Imara 2. Gharraf canal 3. Euphrates entering marshes 4. Old Junction of Euphrates and Tigris at Qurna 5. Shatt al Arab.

The other standing waters are shallow inundations, formed by river action, many of them are permanent and of great though changing extent. Al-Hamed assessed the area occupied by these riverine waters at 20 000 km²; this is an outstanding feature of Mesopotamian Iraq. Some of these 'hors' or 'haurs' as they are called in Iraq, are extensive e.g. Habbaniya 430 km², Thartar 2700 km², Dibbis 1850 km², Hammar 2500 km². This last 'lake' is part of the great southern marshes, extending over 10 000 to 15 000 km² around the final stretches of the Euphrates and the Tigris. An archipelago of extensive reeds is transected by open waters and communication channels; here rivers disperse again and recollect, to unite at Qurna until the late 19th century A.D. and again more recently via the Hor Hammar near Basra (*Fig. 17*). This latter mouth of the Euphrates is indeed a recent phenomenon; historical (Arab) sources mention the formation of the extensive area of the Hammar 'lake' in about 600 A.D. The old channel of the Euphrates is that one towards Qurna, now apparently active during the flood season. The whole area near the delta of the joint river is under continuous changes as a result of river action and the influence of the tidal sea Gulf. Significantly these regions were called in antiquity 'sealands'. The space photograph illustrates the complexity of marshes and rivers. The much discussed problem of the delta will be reviewed later under Hydrology and Sedimentation.

Under the influence of the hot summers many standing waters and marshes undergo drastic changes and in some cases dry out completely and revert to desert conditions.

The Shatt al Arab

There seems to be no doubt that this joint outflow into the sea is a 'recent' formation of the last 5000 years. If we accept the archaeologists' view, reinforced by modern geomorphological opinions, some Sumerican cities were on the sea shores (*see Fig. 23*); and historical later sources locate Basra and especially the Iranian port of Abadan near the gulf. Today Basra is 140 km inland and Abadan about 50 km. The stretch from Qurna to the second Euphrates junction at Basra is 60 km long making the total length of the Shatt about 200 km. The width varies from 400 to 1500 m, its depth from 7.5 to 12.5 m, though besides deeper 'holes' shallows exist; the latter have to be dredged in this important commercial channel. This is necessary because of the big sediment load brought down from the Iranian plateau by the Karun river, which enters the Shatt al Arab near the Iranian port of Abadan, south of Basra. The map and, more realistically, a space picture (*Fig. 18*) shows the course of the Karun; it was even in historic times joined by the Kharkey river, which now deviates towards the marshes above Qurna. This is another example of the continuous changes in the Mesopotamian alluvial plain. The load of sediment of the Karun has blocked the flow of the Tigris and Euphrates and contributed largely to the delta formation. Water levels in the Shatt are under the influence of sea tides, from 3 m near the mouth to 0.5 m at Basra; the influence of the tides is felt about 140 km inland with a

Fig. 18. The Course of the Karun river descending from the southern Iranian plateau. This river with a big sediment load is mainly responsible for the formation and advance of the Delta into the Persian (Arabian) Gulf and terrestrialisation of ancient ports. Its previous tributary, the Kharkey is now deviated towards the great marsh of Hor (Haur) Hawiza, as dicernible in the middle of the picture. The Tigris is visible to the right of the marsh. 1. Karun river 2. Kharkey river 3. Hor Hawiza (Landsat.)

penetration of marine organisms upstream. The Khor Zubair visible on the space photograph (*Figs. 25, 26*) is a recent formation, apparently saline.

This concludes a survey of the river system. The Handbook of Iraq (1944) contains detailed descriptions of the component stretches of the two main Mesopotamian rivers with remarks on width, depth, courses, settlements new and old, riverine basins and dams. A number of painstaking sketch maps and photographs make this book an indispensable guide in spite of being compiled nearly 40 years ago.

Hydrology

The essence of rivers is the transport of water and, as will be discussed in later pages, the transport of sediments. Any biologist dealing with running waters, big or small, should be aware of the importance of hydrology for biological events.

In the case of the Mesopotamian rivers both functions are of utmost importance. All rivers in Mesopotamian Iraq are under the influence of seasonal floods; these are caused by winter rains and snow melt mostly outside Iraq's political territory. Iraq lives on foreign water supply like Egypt.

The spring floods, from February until May with considerable fluctuations, cause first of all an often violent rise in *discharges*: further, they result in rises of the hydraulic force and the currents, carrying sediments; thirdly, floods cause breaches of normal banks and dykes with large scale inundations, especially in years of exceptionally high discharges.

There is a vast amount of data on discharges in the appropriate government archives and publications of Syria, affected by the Euphrates, and in Iraq dealing with both rivers and the Tigris tributaries. Only some data are mentioned here, as far as necessary for understanding the nature of the river system, these are taken mostly from limnological sources.

Al-Hamed (1966) has quoted data for 1957–8 for the two main rivers and two tributaries at stations along the rivers, during high and low flow.

The table shows the great differences of water volume during the flood season and the low water months, the influence on the width and depth of the rivers and differences between both main rivers. Gauge readings published by

Locality	Date	Discharge in cu m/sec	Maximum depth m	Surface width m
Tigris				
Mosul	March 57	4667	8.6	284
	Sept. 57	180	4.3	205
Baghdad	March 57	4099	16.3	204
	Oct. 57	283	10.6	178
Kut	May 57	4491	13.8	350
	Nov. 58	63	4.7	182
Amara	March 57	794	8.0	158
	Nov. 58	42	4.0	138
Euphrates				
Hit	May 57	3120	7.9	285
	Sept. 57	232	5.0	255
Hindiya	May 57	1783	7.1	218
	Nov. 58	159	5.1	201
Nasiriyah	June 57	1309	14.5	152
	Aug. 58	57	10.3	107
Gr. Zab				
Eski Kalak	April 58	902	8.7	176
	Nov. 58	64	3.9	115
Diyala				
Local. not	April 57	867	3.6	142
given	July 58	31	2.2	94

Ionides (1937) from 1906 to 1932, taken at Ramadi on the Euphrates and at Baghdad on the Tigris (*Fig. 19*), give a balanced view of the flood and low regime over 26 years. Rains in the mountains start a rise in water volumes in late autumn.

Sahaf has contributed to our knowledge of the regime and water characteristics of the Euphrates and the Tigris in a paper, written curiously in Russian (1975). A graph from his paper, in adapted form, is reproduced here

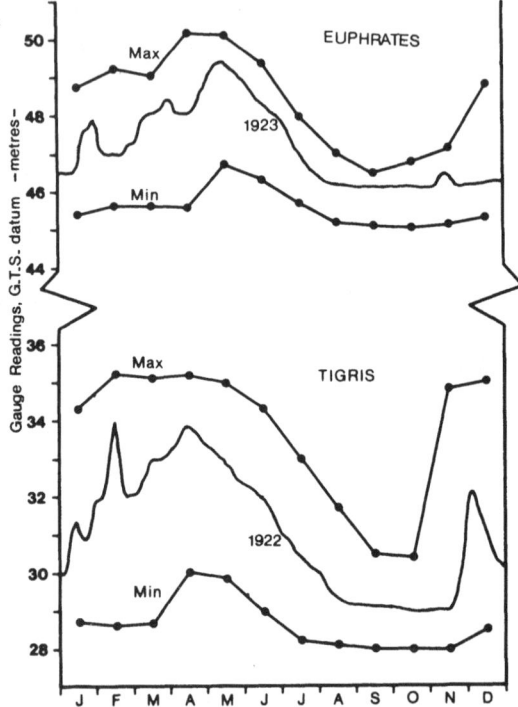

Fig. 19. Discharges of Euphrates at Ramadi and Tigris at Baghdad; averages of 26 years (1906–1932), minima and maxima. A single year is inserted for comparison. (Adapted from Ionides, 1937.)

(*Fig. 20*). It shows the discharges, water temperatures and mineral content of the Tigris at Baghdad throughout 1949. In the same paper Sahaf points out the great changes of the total annual discharges in three consecutive years: in the Euphrates (at Hit or Falluja) 1967 45 km^3, 1968 53 km^3 and in 1969 63 km^3; figures for the Tigris at Baghdad are 46 km^3, 56 km^3 and 93 km^3. Such changes make figures for single years episodical and underline the violence of fluctuations, responsible for flood disasters over many centuries. At Baghdad river levels rose in a recent year by 6 m in few days; in the Euphrates sudden rises of 3 m have been recorded.

Currents

The velocity of water flow is a factor of utmost importance in the ecological make-up of rivers. It governs the carrying and deposition of sediments, the force acting on the river-bed by erosion and on the shape of the river course

50

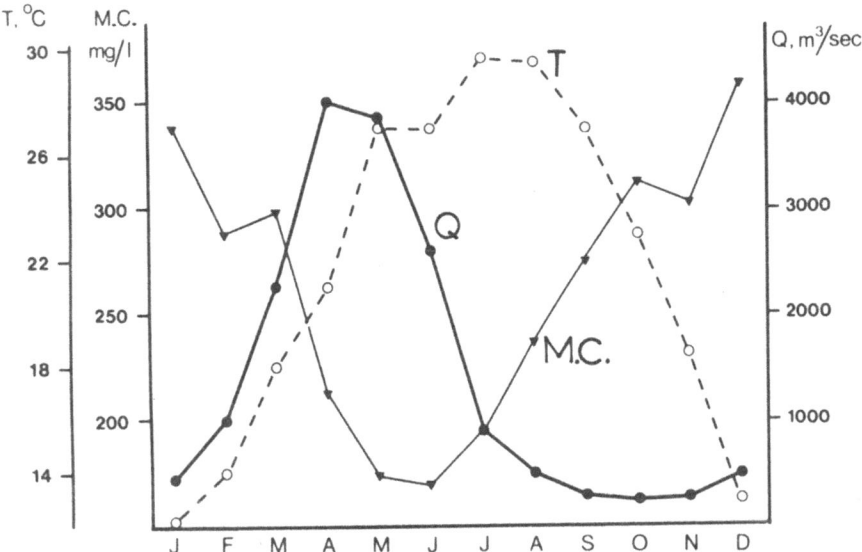

Fig. 20. Seasonal variation of discharge (Q) and salt(mineral)–constant(MC) in the Tigris at Baghdad during 1949; also water temperatures (T). (From Sahaf, 1975, altered.)

or its dissipation, to name some abiotic results. The biological effects are plankton development, distribution of bottom-fauna and others.

Currents depend on slope of river courses, obstacles to discharge and its volume. Too often these obvious decisive factors have been neglected in 'limnological' writings.

The connection between discharges and currents in 1975–76 in the lower Tigris have been given in a paper by Saad and Antoine (1978a). see p. 85.

	Discharges m³/sec and dates				Currents m/sec and dates			
Locality	maxima		minima		maxima		minima	
Kut barrage	May 76	2450	Nov. 75	175	April 76	1.15	Sept. 76	0.31
Amara	Apr. 76	821	Nov. 75	84	April 76	0.90	Nov. 75	0.18
Qalat	May 76	58	Dec. 75	8.2	Dec. 75	0.52	Sept. 76	0.32

Although these figures pertain only to a stretch of the Tigris, they illustrate the general principle of the increase of currents through the increase of water volume by the seasonal flood. A look at the general map and the description of this stretch (Handbook of Iraq 1944) explains the anomaly of greater current at Qalat Salih; there are narrows of the river-bed there, through which the water has to rush. Similarly the relatively high velocity of 0.9 m/sec in the Euphrates on breaking through the Bishri gorge of the Arabian desert shield in Syria is an outcome of river bed configuration. Downstream from the gorge the Euphrates enters the alluvial plain of its own making and begins to meander tortuously, a sure sign of slackening of current (*see Fig. 14*). Such stretches can be found in several parts of the river system, in close inter-dependence of water velocity and sedimentation.

51

Sediments and their deposition

This is one of the outstanding faculties of rivers and the Mesopotamian rivers are the most eloquent example; they have created the plain of historical significance.

From the time of the mountain building Tertiary, enclosing the synclinal basin of low land, erosion products have began to fill the present alluvium of Mesopotamia. The hydraulic force of many streams joined into rivers carried sediments down and sorting them out according to the strength of water volume and currents.

On the basis of observations in Austrian water courses, Einsele (1957) has assessed the carrying capacity of running water. At a speed of 1.70 m/sec and above, running waters scour the river bed and its valley, carry bedload, coarse gravel and shingle downstream and depose these in the upper courses. At 0.50 m/sec sand begins to fall out; at 0.2 m/sec loam and silt are deposited. With possibly slight modifications these rules are universally valid. In Mesopotamia the hydraulic force is governed by the seasonal flood and most suspended matter is carried and deposited during this time.

Al-Hamed (1966) has assessed the suspended matter – equivalent during the 1958–9 flood of the Tigris at Baghdad at 2800 mg/l and 6920 mg/l in the Euphrates at Hit; this fell in summer to about 28–45 mg/l in the Tigris and 46–86 mg/l in the Euphrates. Such data for a single season give only incidental values as the rivers fluctuate widely from year to year, as recorded by Sahaf (1975). Nevertheless, they do indicate the importance of the flood season. Cressey (1958, Middle East Journal 12: 448–460) has commented on the increase of sediments in the Shatt al Arab by the Karun river; this is important, because of the Euphrates and Tigris sediments 90 per cent are deposited in the middle and lower stretches and only 10 per cent reach the Shatt. Depressions, inundation basins and marshes act as silt tanks. Illustrations like the view of the Khabur river at Zakho in the Kurdish foothills and sandbanks at Baghdad show the deposition of gravel and sand (*Figs. 21 and 22*). Space photographs show the interception of both great rivers by marshes (*see Fig. 17*).

The enormous sediments of the Mesopotamian twin rivers, said to be five times as big as those of the Nile, have covered the sea deposits of the Miocene period, salt gypsum and chalk with a thick layer. The marine deposits are activated and influence the groundwater by irrigation practices and rain in the north. The map of soils adapted from Guest (1966) shows the saline nature of most of the alluvial basin (*see Fig. 3*). Even in the relatively short period of historical times, the silts of the middle and lower stretches of the Mesopotamian rivers have covered ancient sites with a layer up to 10 m thick. But the greatest phenomenon is the formation of the delta.

The delta, advance and retreat

Six maps in the *Handbook of Iraq and the Persian Gulf* (1944) tentatively depict the advance and retreat of the ancient 'sealands', ranging in time from

Fig. 21. Transport of Erosion Products; the Khabour mountain stream at Zakho, northern Iraq near Turkish border, with stone and gravel deposits. (Courtesy of Iraq Petroleum Co., London.)

3000 B.C. to 1850 A.D. (*Fig. 23*). They are based on historical sources, covering the last stages of delta development. The earliest map shows the towns of Eridu, Ur, Larsa, Kutalla Lagash lying on an arc of the sea shore, as ports of trade, documented by installations and clay tablets, recording imports. Maps dated approximately from 1500 to about 800 B.C. show the advance of the delta, the gradual separation of the ancient cities from the gulf shore, the courses of the Mesopotamian rivers and their interlocking by side arms. The most spectacular advance is noted for the Karun system, consisting of the Karun and the Kharkey then united and spilling their load of sediments from the Persian plateau into the Gulf (*see Fig. 18*), gradually cutting off the northern part. The plug of sediments prevented the entry of Mesopotamian sediments into the sea, and the basin north of the Karun mouth was filled gradually by silt islands and marshes. From the map of about 800–1200 A.D. it seems that the Euphrates and the Tigris were then forced to join and form the Shatt al Arab. By 1850 the last map shows nearly present conditions. The straight course of the Tigris from Kut al-Imara downstream, the deflection of the Euphrates eastwards towards Qurna and the Shatt al Arab are therefore regarded as relatively recent formations. This seems to be confirmed by the historical facts that Abadan was a port on the open sea in the 10th century A.D. and that the Karun, now separated from its once tributary the Kharkey, enters the Shat al-Arab near Abadan.

The whole problem of the advance of the delta has been in the last decades

Fig. 22. Tigris Meanders showing deposition of sandbanks and silt with decrease of current force. (Aero-Films, London.)

the subject of controversy. Based on older sources the archaeologist Seton Lloyd (1955) drew a map of Mesopotamia with the sea gulf reaching inland to Tikrit north of Baghdad. Such extension of the gulf was denied by Lees and Falcon (1952), no transgression of the sea occurred, and rather a subsidence of the sea of the Persian Gulf took place. The advance of the delta by sedimentation was disputed because these would have filled the swamps surrounding the final stretches of the rivers long ago, if not for the lowering of the sea bed. To this Ionides, who studied the regime of the twin rivers (1937) replied in 1955, that the sediments could not fill the southern swamps because they are deposited over a large area of the river valleys, and that therefore this argument of Lees and Falcon is void.

In 1975 C. E. Larsen took up the delta problem again and based his conclusions on a number of recent studies, and a balanced view is presented. Subsidence of the sea bed is not accepted, some arguments of Lees and Falcon are modified. A map by De Morgan (Delegation en Perse, Paris, 1900) is reproduced and two historical sources are quoted, one a chronicle by Sennacherib 696 B.C., and one from 325 B.C., based on the journal of Nearchus. On this map the sea reached to south of Amara engulfing Basra and the present junction of the Euphrates and Tigris. Marine deposits lie under alluvial silts showing the interplay of two opposing forces, the sea and

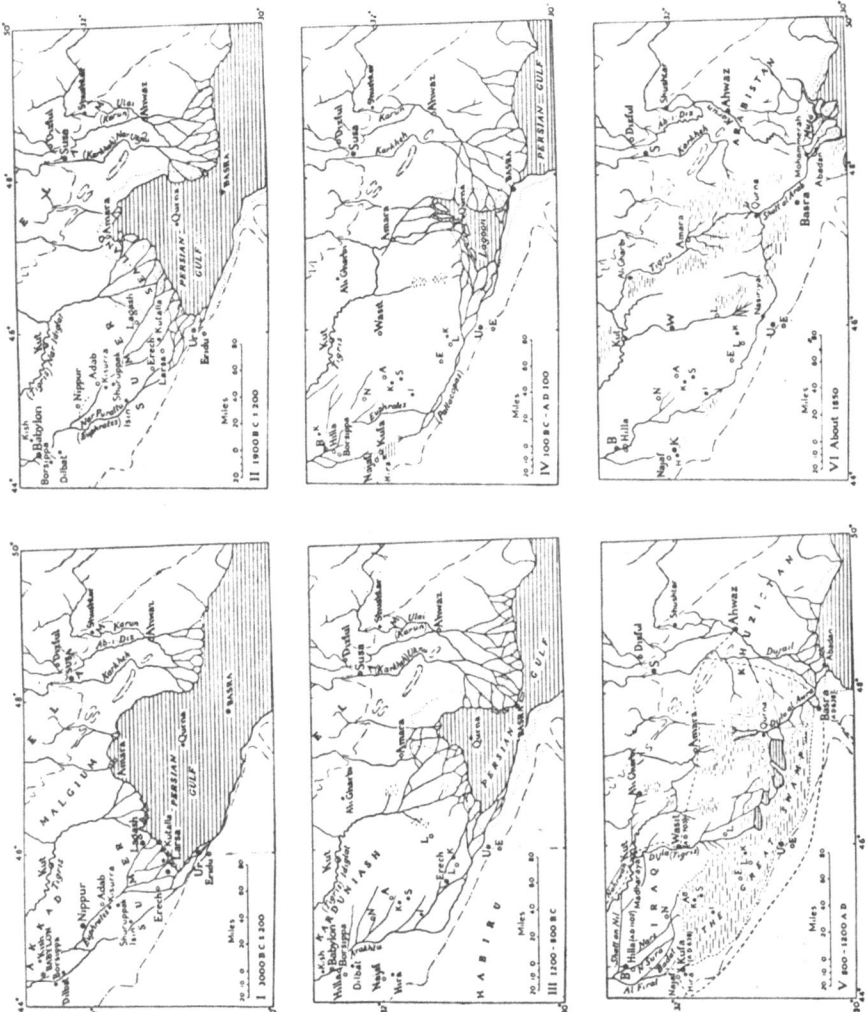

Fig. 23. Development of Delta in Historical Times. Six maps show the probable advance to the sea shore from 3 000 B.C. onwards; the role of the Kharun river is demonstrated. Explanation of human settlements are taken from source book: Place names in sloping type are of towns not yet founded at the time; those in upright type are of towns in being at the time stated on the map; places shown by initials have either disappeared or are inhabited now only by few persons. Land approximately above 500 feet is shown stippled. (From Handbook of Iraq, 1944.)

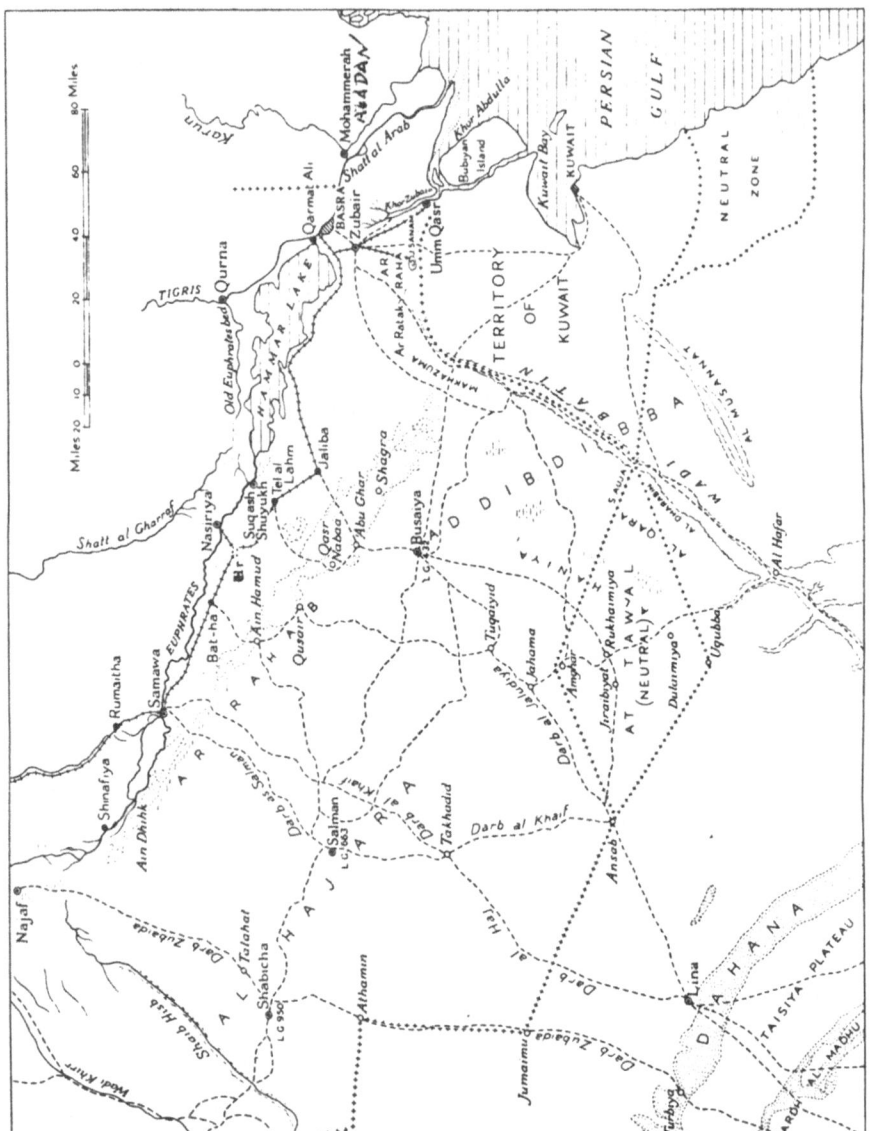

Fig. 24. Delta Area, sketch map showing present configuration of localities, rivers and delta. (From Handbook of Iraq, 1944.)

the force of the rivers. Larsen maintains that the delta has advanced during historic times about 150 to 180 km. Apparently the process goes on and calculations assess the rate of advance seawards at 2 miles, over 3 km, per century. Larsen in his conclusions sees as an important factor the forced adjustment of the rivers and delta configuration to Holocene sea levels, explaining the advance and retreats of the delta; the claims that ancient Sumerian sites were sea ports are justified. Transport of sediments especially by the Karun river together with wind-blown eolian dusts and sand have covered some sites, testifying to this process (*Fig. 24*).

To illustrate the complexity of the southern ancient 'sealands' and their replacement by marshes and the changing image of the delta itself, I reproduce space photographs (*Figs. 25 and 26*). These are impressive and give an overall orientation but require explanation and detailed ground maps; the newest were not available to me. Though some excellent sketch maps in the *Handbook of Iraq and the Persian Gulf* (1944) were made nearly 40 years ago, they are still of great value, showing also changes in the arrangement of the land; these changes continue incessantly (*Fig. 24*).

The space views of the delta were taken in an interval of 5 years, July 1973 and September 1978, during the low state of the rivers; the 1973 picture was taken in a 'softer' wave band. Both comprise the Shatt al-Arab from Qurna to the entry into the sea, though a slight shift in the area is noticeable. The sites of some landmarks have been added by me. The general impression of recent formation is convincing, the entry of sediments into the sea is visible on the 1973 picture (*Fig. 25*) and, though admittedly slight, a change is noticeable to the careful observer. The continuation of delta formations of smaller streams, east of the main exit of the Shatt al-Arab is of interest in view of recent discoveries of ancient sites between the Karun river and the adjoining marshes (*see Fig. 26*). A detailed description is added to each picture. More difficult for detailed interpretation are the southern marshes around and above Qurna represented by another space photograph (*see Fig. 17*).

A previous photograph (*Fig. 18*) shows the country east of the great Hor Hawiza area of marshes and open waters at the low season (July 1977) with the Kharkey deviating from its historical junction with the Karun; now the Kharkey has turned west and feeds the Hor Hawiza with its waters, another example of the continuous shifts of the river system. As previously mentioned both these rivers come from the Iranian plateau. Note the tortuous course of the Karun, which plays such a decisive role in the Shatt al-Arab and the formation of the delta since the earliest historical times.

The Mesopotamian rivers as life arteries for man

Herodotus was a shrewd though late observer of man's life in upper Mesopotamia by the 4th century B.C. The motto used by me attests to the wealth of Assyria and the climatic necessity of irrigation. The location of ancient sites shown on a map is incomplete but eloquent. All ancient sites are on the rivers or within the former reaches of them (*see Fig. 7*). This has continued through the last 5 or 6 thousands of years until now (*see Fig. 7a*).

57

Figs. 25 & 26. Two space photographs of the Shatt al Arab and the Delta into the Persian (Arab) Gulf, taken on 9 July 1973 (Fig. 25) and 29 September 1978 (Fig. 26). Both pictures differ slightly in area; Fig. 26 shows marshes and their drainage channels east of the Delta including the entry of the Karun river. Fig. 25 was taken in a 'softer' wave band and shows the flow of sediment into the Gulf. Both pictures reveal for the discerning eye slight differences even after the short interval of five years. Sites numbered: 1. Old junction of rivers at Qurna; 2. The new entry of Euphrates into the Shatt al Arab near Basra; 3. Entry of Kharun with Mohammera and Ibadan on the Iranian side; 4. The brakish Khor Zubeir, its link with the Hammar lake is faintly visible; 5. Sediment flow into Arabian Gulf. Note the grey areas of cities, cultivation and palm-groves along the Shatt al Arab. (Landsat.)

Prehistoric sites, mentioned in the chapter on Palaeo-ecology, are also marked: there primitive man could survive with adequate rain and good vegetation. His descent to the plain was hazardous and could only be achieved by better management of nature.

Thus irrigation and domestication of plants came into being. Herodotus contrasted basin irrigation, which he knew from his Egyptian travels with the lifting of water by man-operated pumps in Babylon in the 4th century (see

Fig. 26. For legend see page 58.

Annexe). Flood cultivation has been also practised in Mesopotamia, but historical records exist to document the great interest of the ancient city states and later empires in the execution and preservation of irrigation channels, weirs and other installations to ensure water distribution as basis for food production. A dense network developed through the centuries apparently reaching a peak in Abbasid times in the 8th and 9th century A.D. Many of the old irrigation works have almost disappeared and are traced by archaeologists and historians. Some principal channels are said to be ancient courses of rivers, like the Gharaf near Kut al-Imara on the Tigris. River-stretches splitting into a diffuse array of small waters were used also as sites for cultivation. There was through the ages the threat of disastrous flooding and from early times riverine depression basins were used for flood water storage as in the Habbaniya basin, the Dibbis and the Thirtar (Tharthar). On the Tigris the river banks, formed by the big sediment load, had to be continuously reinforced and dyke inspection was a high priority of river surveillance.

Fig. 27. Barrage at Kut el Imara (Amara) on the Tigris before the river and its Gharaf channel enters the southern marshes. Built by Sir William Wilcocks, the barrage regulates the flow distribution of irrigation water. Compare with no. 17. (Courtesy of Iraq Petroleum Co., London.)

Nowadays barrages, weirs and reservoirs have been and are built for a more efficient management of water distribution. The Hindiya barrage regulates the distribution of the Euphrates which has split here into two main channels, the eastern Hilla branch and the western Qufa. By 600 A.D. Arab chroniclers reported the detrimental effects on the less supplied area; now the barrage has solved this problem. Similarly at Kut al Imara a barrage distributes water into the Gharaf channel and the Tigris below the dam (*Fig. 27*).

Two artificial reservoirs have been built in the Kurdish mountains at Dokan on the wild Lesser Zab in 1960 (*see Fig. 16*), and one as part of the Diyala development project at Derbendi-Khan in 1961. Both are deep water bodies in mountain valleys and their main purpose seems to be for irrigation.

A major project is under way to drain a large part of the southern marshes around Qurna. Like the Jonglei scheme on the upper White Nile this will cause ecological changes of unknown extent.

A number of projects are under way in some parts of Iraq, sponsored by the government, re-afforestation in the Kurdish foothills and mountains, use of the humid steppe lands for range deployment and others to ameliorate the ravages of the past.

The river system as channels for communication

Rivers allow human communication, trade and cultural exchanges. The Babylonian frieze reproduced earlier (*see Fig. 10*), is one of the many bas-reliefs existing. The book by Parrot (1961) contains a splendid series of illustrations which allow a glimpse into the importance of river traffic in Mesopotamia. All important settlements had to lie on the rivers and main channels and boat building reached a high level of craftsmanship. This tradition survives in full today on the rivers, though of course replaced largely by power-driven craft. The 'Marsh Arabs' use boats of surprisingly elegant and efficient design, as described by Thesiger (1964) and Young (1977). The extraordinary coracle, pictured on the Babylonian bas-relief (*see Fig. 10*) is still in use after so many centuries. Herodotus in his Histories of the 4th century was so impressed that he gave a detailed description (see Annexe). His notes reveal vine cultivation in the north and brisk trade along the river; the shortage of wood in mid-Mesopotamia is also confirmed by log-rafting from the forests in the north. Other goods were also traded by river transport. Navigation by sail was difficult; but larger boats called locally 'mahailas' and 'safinas' are now in use, though again being replaced by power-craft. Ferries have existed since ancient times, understandable in a country transected by so many waterways. The traveller Heyerdahl has recently proved the possibility of seafare by building a reed boat of great proportions and sailing from the Persian Gulf to the Red Sea; thus he confirmed that from Sumerian times onwards a communication could have existed to other sea ports.

This is commerce; other fields of river communication include 'exported' creeds, people of different races and languages all along the rivers. Apparently there were connections between the Al-Ubaid people with the settlements of Halaf and Hassuna even before Sumerian times. Rivers are life arteries.

6 Water characteristics

J. F. Talling

The regional factors

Water characteristics in Mesopotamia, as in other regions, are determined by an array of influences which reflect the prevailing climate, the substrata of soils, rocks, and ground-water, biological activity in the waters themselves, and human manipulations. The strongly seasonal climate enforces changes in time, especially of such physical characteristics as temperature, turbidity, and discharge. It also interacts with the substrata to yield a considerable range of chemical conditions. The most influential human manipulations are connected with hydrology and agriculture, rather than enriching or polluting inputs.

Under the influence of a climate with winter rains and hot rainless summers, monthly evaporation rates from open water may vary between 26–89 mm from November–March and rise to 178–321 mm in May–September; after correction of data the annual evaporation varies between 1670 and 1710 mm. These figures apply to the Syrian sector of the Euphrates just before entering Iraq, given in Brinkmann's paper (1967) on irrigation management in that region. Further south, for the Tigris near Baghdad, Saad & Antoine (1978a) quote a variability between 53.1 mm/month in December and a very high value of 752 mm/month in July. For Basrah, Saad (1978b) cites a range between 55 mm/month in January and 300 mm/month in August. Crops increase evapo-transpiration up to 3000 mm per year. These values may be compared with others for rainfall. Rain over Mesopotamia falls in winter; it varies from 300 to 1200 mm/year in the hills and mountains, but in most of Iraq and Syria is below 200 mm/year. Thus rivers are main sources of water, without which the arid areas would spread.

Rain has some effect on the vegetation, but for economic food production irrigation has to be used. Here the mutual impact of soils on the irrigation waters, and *vice versa*, starts. All arid zone soils tend to be saline in varying degree, depending on solid substrata and the nature of ground-waters; in fields left fallow salts are sucked to the surface by evaporation and may form visible encrustations. Saline soils have to be leached by river water; canals for this drainage, if created, carry large amounts of salts into the rivers.

To this must be added the large areas of flood basins along both main rivers in Iraq, filled by the floods and diminishing during the summer. As irrigation intensity increases downstream, it is understandable that the general tendency is for increased salt contents downstream, augmented by geological factors, and modified by seasonal floods. This complicated chemical regime causes considerable difficulties for land management. Some repercussions of the problem of soil salination can already be found in the ancient riverine civilizations of Babylon and Assur.

River water arriving in Mesopotamia undergoes changes in chemical composition by the extraordinary contact with the land – here, much more than in the Nile valley, a paramount factor in limnology. The ground-waters are under the influence of regional soils and their substrata, limestone ($CaCO_3$) and gypsum ($CaSO_4$) in the north and NaCl mainly in the middle stretches of both rivers. The Euphrates is under stronger influence of the arid steppe-desert than is the Tigris flowing along the Zagros mountains and receiving four important tributaries from the mountains. As soon as the main rivers enter the great alluvial basin their chemistry changes. In this ground-waters are particularly influential.

Ground-waters are the result of penetration, uplift due to surface evaporation, and capillary powers. They represent solutions of salts, entering the rivers via drainage channels, seepage, inundations etc. Their chemical composition is represented in Table 1 from Al-Sahaf (1975). From this table the following conclusions emerge:

(1) there is a sharp increase towards the south, in the ground-water basins surrounding the Tigris, of Na^+, Mg^{2+}, Cl^- and SO_4^{2-}, with a slight decrease in Ca^{2+}
(2) the Euphrates has higher values of the above, especially in Na^+ and Cl^- as recorded at one site mid-course
(3) the Shatt al-Arab ground-water (at one station) has much lower values, probably because of the marshes in which the soils are not subject to leaching effects.

The transfer of solutes from land to water also reflects the prevalence of readily soluble minerals. Thus limestone is widespread in the mountain ranges of the north and east, where tributaries originate. From the opposite direction, past transgressions of the sea have left a residue of marine salts in the sediments of the lower plains.

Temperature

Whether flowing or standing, surface waters of Mesopotamia show considerable seasonal ranges of temperature which reflect the geographical latitude and associated regime of solar radiation. Additional influences are exerted by the higher altitudes to the north, and by the moderating water-mass of the Gulf to the south. A near-constancy of temperature with season can only be found at springs where underground water emerges. A remarkable example is the massive Serchinar spring near Sulaimaniyah, Iraqi Kurdistan, studied in some detail by Maulood & Hinton (1978), where temperatures lie at 17.7 ± 1°C throughout the year.

Patterns of seasonal variation over the Euphrates–Tigris system are illustrated in Fig. 20 (p. 52) and Table 2. For the Tigris at Baghdad, Al-Hamed (1966) cites a range of monthly mean temperatures in surface water during 1958 of between 8.5°C (January) and 31.4°C (August). For Amara, further south, the range in 1958 is slightly less, 10–29.3°C. However,

Table 1. Chemical composition of ground waters in the Tigris and Euphrates basins in 1972, at Mosul, Baghdad and El Suweira – Musaiyib – Shatt al-Arab at El Zubeir (from Al-Sahaf 1975).

River	Site	pH	Na^+	K^+	Ca^{2+}	Mg^{2+}	Cl^-	SO_4^{2-}	HCO_3^-	CO_3^{2-}	Sum, cations + anions
Tigris	Mosul (El-Gaslani)	7.6	275	—	512	160	93	2110	248	—	3460
			11.9	—	25.6	13.2	2.62	43.9	4.06	—	101.3
	Baghdad (Abu-Greib)	7.92	2990	23.5	476	961	2638	7589	104	0.0	15 881.5
			130	0.6	23.8	78.8	74.3	158	1.7	0.0	467.2
	El-Suweira	7.65	4234	57.1	432	1449	2283	12 248	384	3	21 090
			186.7	1.46	21.6	118.8	64.3	225	6.3	0.1	654.4
Euphrates	Musaiyib	7.55	7475	52.8	756	922	12 425	4236	92	0.0	15 958.8
			325	1.35	37.8	75.6	350	88.2	1.5	0.0	879.4
Shatt al-Arab	El-Zubeir	7.8	235	—	549	226	—	1451	150	—	4550
			10.2	—	27.4	18.6	—	30.2	2.46	—	88.9

Data in columns: (1) mg/l, (2) m equiv/l.

65

Table 2. Seasonal variation of surface water temperature at various localities in Iraq, including the rivers Tigris and Euphrates (as mean monthly values in 1958, from Al-Hamed 1966), the Shatt al-Arab (in 1973, from Arndt & Al-Saadi 1975), and L. Habbaniyah (in 1957–1958, from Al-Kaisi 1964).

| | Tigris at | | Euphrates at | L. Habbaniyah | Shat al-Arab |
	Baghdad	Amarah	Samawa		at Basrah
January	8.5	10	9.2	12	14.3
February	12	11	12	12	15.1
March	16.7	16	16.7	15	—
April	20.6	19	20.6	18	—
May	26.3	23	25.3	25	27.3–30.4
June	28.7	25	26.6	27	—
July	29.4	26.8	26.7	28	—
August	31.4	29.3	29.3	28	31.8–32.5
September	30.0	27.4	28.1	25	28.6
October	24	23.7	23.7	24	26.7
November	15.7	18.5	16.9	17	—
December	12.5	12.6	12.3	12	14.8

it is not greatly different from that illustrated by Al-Sahaf (1975) for Baghdad during a later year, or from that given by Al-Hamed for the Euphrates at Samawa during 1958. All these seasonal changes are marked by rapid warming in March and April, and rapid cooling in October and November. Water temperatures are in excess of 20°C for about 7 months of the year. An extensive description of longitudinal changes in temperature along both rivers is apparently lacking, but Saad & Antoine (1978a) describe the situation during 1975–1976 in the lower Tigris between Baghdad and Qurna. There, during the cooler months of January and April, temperatures tend to increase southwards. This trend is lacking in the hotter months of July and October. At some stations there were small differences of temperature with depth, indicative of a slight (diurnal?) development of thermal stratification.

In the Shatt al-Arab, the vertical stratification of temperature was very small (<1°C) or absent at the stations worked in April 1974 by Saad & Kell (1975) and Kell & Saad (1975), and in February 1978 by Maulood et al. (1979). During these two periods, the ranges encountered were respectively 25.0–25.3° and 15.5–17.0°C. The seasonal extremes, measured at Basrah by Arndt & Al-Saadi (1975), were 14–15°C in January and 32°C in August. Even at the higher seasonal temperatures, a marked thermal stratification was lacking in this flowing and relatively shallow water (Mohammad 1965b, Arndt & Al-Saadi 1975).

In the *standing waters*, stratification is likely to develop seasonally in the deeper basins, such as the northern reservoir-lakes of Derbendy-Chan and Dokan. In the more numerous shallow basins, often large in area, one would expect only transient and often diurnal forms of thermal stratification, prone to destruction by wind or nocturnal cooling. This lack of stable stratification seems to be borne out by the few published observations on the plain-depression lakes (e.g. L. Habbaniyah: Al-Kaisi 1964) and the southern marsh lakes (e.g. at Ashan-Hammar: Maulood *et al*. 1979). The seasonal variation of temperature in L. Habbaniyah, observed during 1957 by Al-Kaisi, is very similar to that recorded by Al-Hamed (1966) for the adjacent Euphrates during 1958 (Table 2). Al-Hamed lists temperatures measured during February and August 1958, which approximate the seasonal minima and maxima, in a variety of shallow lakes on the Mesopotamian plain (Table 3). The temperature range in February was 15–18°C, and in August 25–31°C.

Table 3. Winter and summer temperatures in several lakes of Iraq, February and August, 1958 (from Al-Hamed 1966).

Lake	Winter (February)	Summer (August)
L. Habbaniyah	15	25
L. Hammar	16	31
Shattra Lakes	16	28
Meshkhab and Shamiyah basins	17	30
Khor Ummal Baqr and Khor Auda	16	29
Qalal Saleh – Qurnah marshes	18	31
Khor Al-Hawisa	17	30

Turbidity and transparency

These optical properties reflect the abundance, or scarcity, or suspended particulate material. Al-Hamed (1966) comments that 'as the river [Tigris and Euphrates] water is very poor in plankton and organic detritus, turbidity is produced almost entirely by silt'. Using a comparative scale of measurement (Jackson turbidimeter), he followed the annual variation of turbidity in the Tigris at Baghdad and the Euphrates at Hit (Table 4). Values are relatively low during the period of slow flow between July and October. They increase to high levels during the following phase of flood water in winter and spring, when the sediment load (Chapter 5) is increased in consequence of heavy rains and melting snow in the mountains, with increased erosion and turbulent transport during river-flow there and in the plains. The seasonal pattern in 1958–1959 shows an abrupt and earlier rise of turbidity in the Tigris at Baghdad, as compared with a more gradual and later developed maximum in the Euphrates at Hit. This conspicuous difference may be linked with the important drainage to the Tigris from the nearby eastern range of mountains.

A related but inverse measure, of *transparency*, is provided by estimations of the depth-limit of visibility of a white disc (Secchi disc). Published values are for the southern reaches, especially the Shatt al-Arab. There Saad & Kell (1975) found values of 50–70 cm during the flood period in April 1974. Although these are low compared with many rivers, they still suggest a considerable shedding of sediment from more turbid waters upstream. Along

67

Table 4. Turbidity, in mg/l of equivalent standard, for the
Tigris and Euphrates rivers, April 1958 to March 1959 (from
Al-Hamed 1966).

Date	Tigris	Euphrates
1958: April	1146	3270
May	598	6920
June	170	2136
July	28	173
August	30	46
September	41	86
October	48	96
November	922	140
December	2407	470
1959: January	2800	1622
February	880	840
March	430	1267

a longitudinal section of the Shatt al-Arab, similar values were recorded in the
same month by Kell & Saad (1975) just before and just after the arrival of
flood water, and in February 1978 by Maulood *et al.* (1979) (*Fig. 28*).
However, there can be a dramatic reduction, e.g. to only 5 cm, after the entry
of the sediment-laden Karun river, followed by a recovery downstream nearer
the sea (Mohammad 1965, Kell & Saad 1975, Saad 1978a, b). Considering
this impact of the Karun river, with its short flow from the mountains in Iran,
the Euphrates and Tigris appear to have shed most of their flood-sediment on
the way south. In standing waters of the marsh region, transparency may
exceed 3 m (Maulood *et al.* 1979). Even in the Shatt al-Arab at Basrah, and in
the main rivers near their junction, values greater than 1 m are known
seasonally (Mohammad 1966, Arndt & Al-Saadi 1975, Saad 1978a); these
indicate minima in the sediment load during low flows. Values of 3 m are
known from the Dokan reservoir (Al-Hamed 1976).

Light penetration increases strongly during the final transition from Shatt
al-Arab to Gulf water. This is apparently the only example in Mesopotamia
documented by photo-electric measurements (Al-Saadi, Arndt & Hussain
1975, *Fig. 2*).

Salinity

Three intercorrelated measures have been used to assess the dissolved salt
contents of Mesopotamian waters. These are: the weight of dissolved material
obtained after evaporation, in mg/l; the electrical conductivity, in μS
($=\mu$mho)/cm at 25°C; and the sum of analyses for the major cations and
anions, in meq/l.

The circumstances which have led to a prevalence of moderately to strongly
saline waters have already been mentioned.

In the *two main rivers*, salinity is inversely related to discharge and tends to

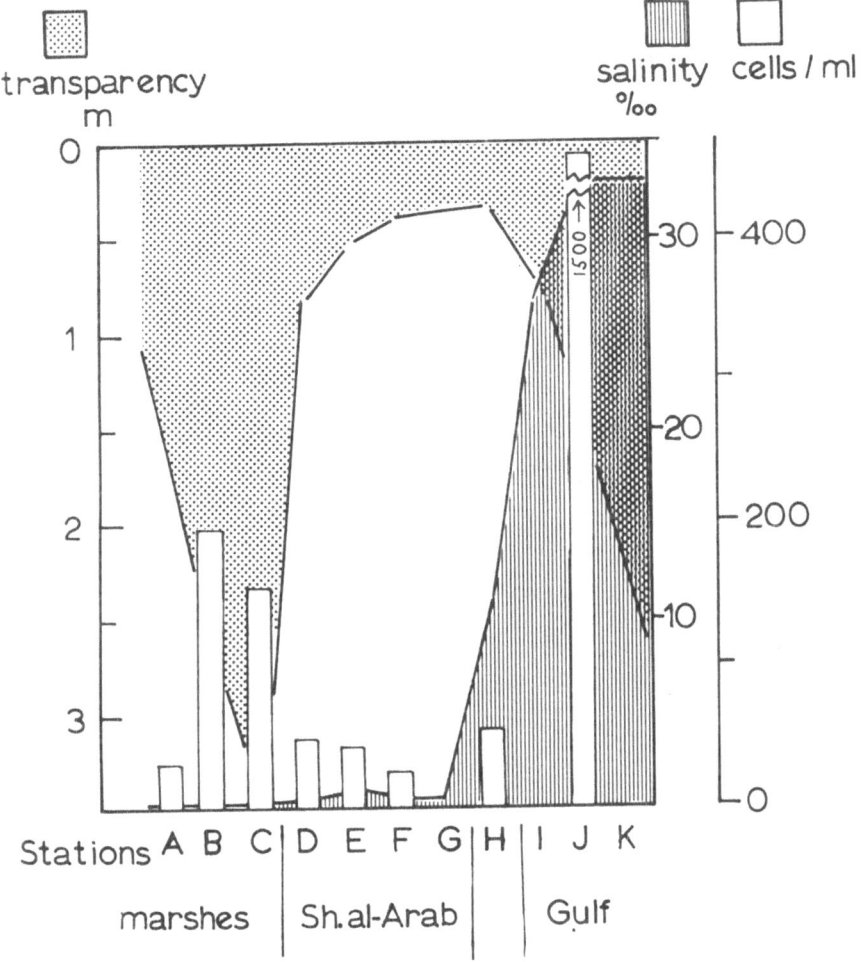

Fig. 28. Transparency (Secchi), salinity and abundance of diatoms in surface water along a downstream section of the Shatt al-Arab from Lake Hammar to Gulf. (From Maulood *et al.* 1975.)

increase during flow southwards. Al-Hamed (1966) illustrates the seasonal variation during 1958 in the mid-plain region, at Baghdad (Tigris) and Samawa (Euphrates). These gravimetric estimations (Table 5) show lower values, under 200 mg/l, during the phases of high discharge, April–June in the Tigris and May–July in the Euphrates. The diluting effect of flood-water is, of course, commonplace in rivers: the Nile provides another example (Talling 1976). After the flood-water is past, back-drainage of water bearing leached salts will contribute to the major rise of salinity (*see Fig. 20*). This is especially marked in the Euphrates, where a maximum salinity above 500 mg/l is recorded in September. Esssentially the same features are described by Al-Sahaf (1975), both from gravimetric estimates (*Fig. 20*) and also using electrical conductivity as the measure of ionic concentration (*Fig. 29*). His longitudinal sections of the two rivers are especially interesting; they show the

69

Table 5. Salinity (mg/l) in the Tigris, Euphrates and Diyala rivers, 1958 (from Al-Hamed 1966).

Date	Tigris[1]	Euphrates[2]	Diyala[3]
January	260	180	170
February	260	180	175
March	220	220	200
April	190	215	270
May	190	184	274
June	190	160	479
July	230	184	346
August	280	450	444
September	300	525	318
October	320	490	354
November	380	375	280
December	370	400	200

[1] at Baghdad.
[2] at Samawa.
[3] near its junction with the Tigris.

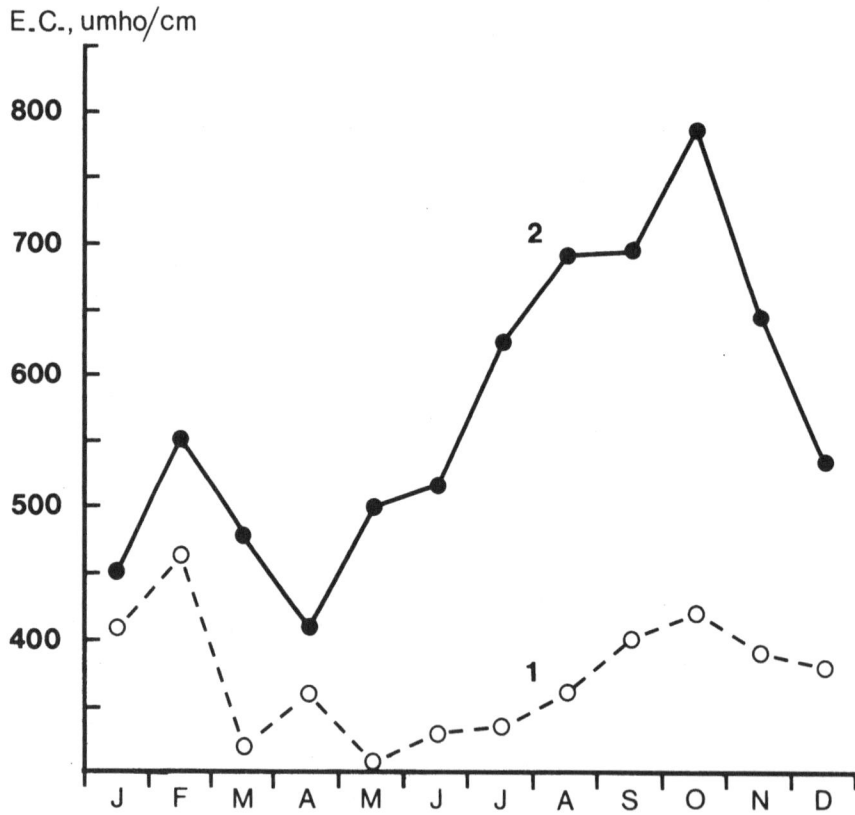

Fig. 29. The seasonal variation of electrical conductivity (E.C. at 25°C) in the Tigris at Samara (1) and the Euphrates at Falluja (2), as average values of several years. (From Al-Sahaf 1975.)

very considerable gain of ionic content during the flow downstream (*Fig. 30*) especially during low water and in the Euphrates (Table 6). There conductivity can reach 1440 μS/cm (at 25°C) – a value far in excess of any known from the Nile.

The *Diyala River* exemplifies features of an influential tributary, with relatively short flow from the mountain headwaters to the final junction with the Tigris near Baghdad. Measures of salt content given by Al-Hamed (1966) (Table 6) and Al-Sahaf (1975) (Table 6) indicate a roughly two-fold seasonal range, with the lowest values in winter before the minimum in the main Tigris.

In the *Shatt al-Arab*, high levels prevail. Not only is there the accumulated salt flux from the north, but a marine influence extends a varying distance up the estuary. The distance is naturally reduced during phases of high river flow (Al-Sahaf 1975; Table 6). Variation with the tidal cycle at two stations is illustrated by Mohammad (1965). The longitudinal distribution of salinity or electrical conductivity has been described by various authors (e.g. Al-Sahaf 1967, Arndt & Al-Saadi 1975, Maulood *et al.* 1979). In the upper reaches, the conductivity is typically about 1000 μS (μmho)/cm at 25°C, followed by

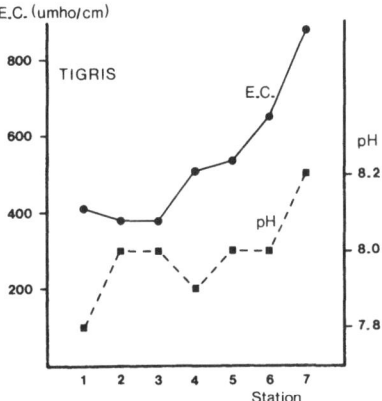

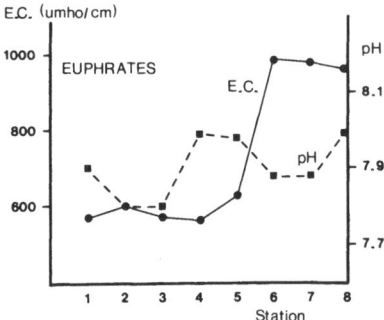

Fig. 30. The longitudinal variation of electrical conductivity (E.C. at 25°C) and of pH at 7 stations (Mosul to Qurna) in the Tigris and in the Euphrates at 8 stations from Qaim to Qurna. Values are averages from several years. (From Al-Sahaf, 1975.)

71

Table 6. Electrical conductivity, in μS ($=\mu$mho)/cm at 25°C, at various stations along the Euphrates, Tigris, Diyala R., and Shatt al-Arab, in relation to water regime (from Al-Sahaf 1975).

Site	Low	Flood	Average
A. Euphrates			
Qaim	731	444	575
Hit	776	470	600
Falluja	681	470	568
Musaiyib	700	462	559
Diwaniya	873	488	630
Samawa	1440	631	984
Nasiriya	1386	640	976
Qurna	1276	765	959
B. Tigris			
Mosul	445	350	404
Fatha	440	352	373
Samarra	446	349	379
Baghdad	674	442	507
Kut-el-Amara	670	441	537
Amara	992	475	650
Qurna	1137	773	880
C. Diyala R.			
Bakuba	1386	638	738
Diyala Bridge	1628	851	1058
D. Shatt al-Arab			
Qurna	1121	720	901
Basrah	1509	944	1182
Fao 1	5064	1053	2802
Fao 2	5400	1064	2947

increase downstream at distances from the sea which vary, for hydrological reasons. Input from an important tributary, the Karun River, also produces an increase (Al-Sahaf 1975).

In the *standing waters*, salt content varies enormously (Table 7). Relatively low values (by Mesopotamian standards) are found in the northern reservoir-lakes of Dokan and Derbendy-Chan, and much higher ones in the depression-basin lakes of the central plain – Habbaniyah, Abu-Dibbis, and Tharthar. The difference between lakes Habbaniyah and Abu-Dibbis has an interesting history, which illustrates the effect of varying water throughput. According to Al-Kaisi (1964), the salinity of Lake Habbaniyah was 3350 mg/l in 1933. The completion in 1950 of an inflow–outflow canal link to the nearby Euphrates allowed a controlled flushing, related to water level in the river, which reduced the salinity to 518–774 mg/l in 1950–1951, 402–642 in 1957–1958, and (Table 7) 200–899 mg/l in 1969–1972. By contrast, the adjacent basin of Abu-Dibbis (connected by a regulated canal to Habbaniyah) has no such through-flow; its very high salt content is due to a combination of evaporation and ready leaching of the soluble minerals abundant in the area. White crusts of salt are conspicuous around this lake.

Table 7. Mineral contents of standing waters, reservoirs and lakes, 1969–1972 (from Al-Sahaf 1975).

Region	Lake	Mineral content, mg/l		
		Maximal	Minimal	Average
Headwater	Dokan	237	152	172
reservoir-lakes	Derbendy-Chan	295	160	222
Plain	Habbaniyah	899	200	644
depression	Tharthar	1674	1360	1511
basins	Abu-Dibbis	9850	5640	6315
Marshes	Hammar	640	384	440

Lakes of the southern marshes are fed directly or by seasonal overspill from the main rivers, and their salinity reflects that of the supply. Values for the biggest, Lake Hammar, are given (from Al-Sahaf) in Table 7.

Major ions

Almost all the ionic content is contributed by four cations (Ca^{2+}, Mg^{2+}, Na^+, K^+) and four anions (HCO_3^-, CO_3^{2-}, Cl^-, SO_4^{2-}). Their absolute concentrations and relative proportions in Mesopotamian waters have been summarized by Al-Sahaf (1975) (Table 8). He also gives, by summation, estimates of total ionic content in meq/l, a variable closely correlated with electrical conductivity. The combined concentrations of HCO_3^- and CO_3^{2-}, in meq/l, approximate the titration alkalinity from which the estimations are presumably derived. Alkalinity values for the Serchinar spring and some southern waters are tabulated directly by Maulood & Hinton (1978) and Maulood *et al.* (1979). Some information on ionic composition is also given by Al-Hamed (1966), but his stated units of mg/l are obviously in error (for meq/l?).

In the *two main rivers*, there are clear trends of ionic proportions from north to south. Near the headwaters Ca^{2+}, Mg^{2+} and HCO_3^- dominate, as would be expected in hill regions with much calcium and magnesium carbonates. During flow south and the acquisition of more solutes, marked increases occur in concentrations of Na^+, Cl^-, and SO_4^{2-} (Table 8). These changes are in accord with the general composition of ground-waters (Table 1) and the ionic contribution from former marine deposits. Concentrations of Ca^{2+} may well be limited by the low solubility product of $CaCO_3$. Along the relatively short tributary of the Diyala River, there are smaller solute gains and shifts in ionic proportions.

These ionic concentrations, from Al-Sahaf, are averages from a two-year period (1967–1969) and will hide important seasonal differences. Some indications of these in the Tigris, at Baghdad and below, can be obtained from the papers of Al-Hamed (1966) and Saad & Antoine (1978a). According to the former, the proportions of Ca^{2+} and $HCO_3^- + CO_3^{2-}$ are higher in the flood season, compared with those of Mg^{2+}, Cl^-, and SO_4^{2-}. The latter paper

73

Table 8. Concentrations of major ions, in meq/1, at various stations along the Euphrates, Tigris, Diyala R., and Shatt al-Arab; average values from various seasons in 1967–9 (from Al-Sahaf 1975).

Site	Na^+	K^+	Ca^{2+}	Mg^{2+}	Σ cations	Cl^-	SO_4^{2-}	HCO_3^-	CO_3^{2-}	Σ anions
A. Euphrates										
Qaim	2.0	0.08	2.8	2.3	7.18	1.6	2.8	2.8	0.3	7.5
Hit	1.7	0.09	2.8	2.3	6.89	1.9	1.8	2.7	0.3	6.7
Falluja	1.6	0.09	2.7	2.1	6.49	1.9	1.7	2.7	0.3	6.6
Musaiyib	1.6	0.09	2.7	2.1	6.49	1.8	1.6	2.4	0.3	6.1
Diwaniya	2.1	0.09	2.6	2.9	7.69	3.2	1.9	2.8	0.3	8.2
Samawa	4.3	0.12	3.3	3.5	11.22	4.9	3.0	2.9	0.3	11.1
Nasiriya	4.1	0.11	3.3	3.3	10.81	4.6	3.1	2.8	0.3	10.8
Qurna	3.7	0.11	3.7	3.6	11.11	3.6	3.3	3.4	0.3	10.6
B. Tigris										
Mosul	0.5	0.06	2.7	1.8	5.06	0.7	1.4	2.9	0.3	5.3
Fatha	0.6	0.05	2.6	1.7	2.95	0.8	1.1	2.9	0.2	5.0
Samarra	0.5	0.06	2.4	1.6	4.56	0.8	1.0	2.6	0.3	4.7
Baghdad	1.4	0.07	2.6	2.2	6.27	1.5	1.6	2.6	0.3	6.0
Kut-el-Amara	1.3	0.06	2.8	2.3	6.46	1.4	1.6	2.7	0.3	6.0
Amara	2.0	0.07	3.4	2.1	7.57	2.1	2.4	2.7	0.3	7.5
Qurna	2.9	0.09	3.8	3.2	9.99	3.0	3.1	3.3	0.3	9.7
C. Diyala R.										
Bakuba	2.3	0.07	3.4	2.7	8.47	2.3	2.9	2.9	0.3	8.4
Diyala Bridge	3.8	0.11	4.1	3.9	11.91	3.6	3.6	3.0	0.3	10.5
D. Shatt al-Arab										
Qurna	3.1	0.11	2.5	3.5	9.21	3.5	3.0	3.4	0.4	10.3
Basra	5.4	0.15	3.4	4.0	12.95	3.8	3.7	3.0	0.3	10.8
Fao	21.4	0.33	4.0	7.2	32.93	23.0	4.9	3.2	0.3	31.4

shows that the general increase of Cl^- downstream is subject to considerable temporal variation.

In the *Shatt al-Arab*, the ions Na^+ and Cl^- rise to dominance – a trend that is naturally enhanced in the lower estuary region. Estimations of Cl^- concentration (chlorosity) have been employed by several authors (Mohammad 1965, Saad & Kell 1975, Kell & Saad 1975, Saad 1978a, b, Maulood *et al.* 1979) as a guide to the admixture of fresh with sea water.

Among the *standing waters*, the major ionic composition of Lake Habbaniyah appears to be broadly similar to that of the Euphrates inflow. For 1962, Al-Kaisi (1964) cites the following concentrations, in meq/l: HCO_3^- 2.0–2.4, Na^+ 1.8–2.0, Ca^{2+} 2.0, Mg^{2+} 1.6–2.2, Cl^- 1.8, SO_4^{2-} 1.4–2.0. A few analyses are available for marshes near Lake Hammar, for which Maulood *et al.* (1979) record the following concentrations (here given as meq/l) in February 1978: Na^+ 2.6–10.1, K^+ 0.06–0.09, Cl^- 2.3–7.9, HCO_3^- 5.1–6.7.

Dissolved gases

The content of *dissolved oxygen* has been measured in several studies, especially of the lower Tigris (Saad & Antoine 1978a) and the Shatt al-Arab (Mohammad 1965, Saad & Kell 1975, Kell & Saad 1975, Al-Saadi *et al.* 1975, Saad 1978b, Maulood *et al.* 1979). None of these provide examples of strong deoxygenation; this is known from the bottom water of the stratified Dokan reservoir (Al-Hamed 1976). In the shallow flowing waters, with sufficient turbulence to prevent prolonged stratification and in the absence of heavy organic inputs (as by local pollution at Mosul: Mahmoud & Ahmad 1979), the entry of oxygen from the atmosphere is apparently sufficient to maintain levels generally above 50 per cent saturation. Exceptions may well exist in the extensive marsh areas with decomposing vegetation, but are not revealed by the few published measurements (Maulood *et al.* 1979) available from such areas. At one station in the lower estuary, where the contact between fresh and salt water produced a density stratification (pycnocline), a low oxygen content of only 2.0–2.4 mg/l was found at 1 m depth and below (Maulood *et al.* 1979).

In three longitudinal sections of the Shatt al-Arab, carried out in February 1965 (Mohammad 1965), April 1974 (Kell & Saad 1975), and February 1978 (Maulood *et al.* 1979), the oxygen concentrations in surface water were usually 50 per cent or more of saturation. With the exception noted above, differences with depth were generally minor. Seasonal variation in the Shatt al-Arab is described by Saad (1978a, b).

Further upstream, another set of four longitudinal sections were made by Saad & Antoine (1978a) during 1975–1976 along the lower Tigris below Baghdad. The concentrations encountered, at all depths and seasons, lay between 62 and 124 per cent; the variation with depth at any station was usually small. As there was no positive relationship with abundance of the phytoplankton, Saad & Antoine concluded that 'the atmosphere is a more important source of DO in the Tigris than photosynthetic activity, and . . . the mixing processes and water temperature are the principal factors affecting its concentration and distribution'. Low winter temperatures obviously influenced the higher mass concentrations found in January and April.

Although direct estimations of the content of free *carbon dioxide* are generally not available, some estimate of the changes in this gas can be made from published values of *pH* and bicarbonate + carbonate alkalinity. For a representative range of alkalinity of 2–4 meq/l, air-equilibrium with respect to CO_2 would be expected to lead to pH values between 8.2 and 8.5 (cf. Talling & Talling 1965, *Fig. 3*). Most values recorded from Mesopotamian river or lake waters are slightly lower, often between 7.0 and 8.0, and so imply a slight excess of free CO_2 relative to air-equilibrium. Examples are provided in the works of Al-Hamed (1966), Al-Sahaf (1975) (*see Fig. 30*), Saad & Antoine (1978a), Saad (1978a, b), Maulood & Hinton (1978), and Maulood *et al.* (1979). Saad & Antoine (1978a) record a few appreciable higher pH values of 8.5–9.25 from the Tigris below Baghdad. Though somewhat irregular, they believed that such high values tended to be associated with phases of phytoplankton abundance and hence CO_2-depletion by photosynthesis.

However, both they and Al-Hamed (1966) comment that the apparent lack of extreme pH values suggests that this factor is unlikely to limit the distribution of aquatic organisms.

Plant nutrients – nitrogen, phosphorus, and silicon

Of obvious importance for biological productivity, the quantities and distribution of these nutrient elements in Mesopotamian waters are very inadequately known.

The most intensive, seasonal study seems to be that of Maulood & Hinton (1978) on the Serchinar spring in Iraqi Kurdistan. Here there is a considerable concentration of nitrate-nitrogen, over 500 μg/l for most of the year, with a maximum in March–April of about 1800 μg/l. Concentrations of phosphate-phosphorus and silicate-silicon are moderate, generally between 6 and 18 μg/1 and 2.5–4.5 mg/l respectively. Though the spring is stenothermal (17.7 ± 1°C), the concentrations of these and many other solutes (e.g. O_2, K^+, SO_4^{2-}, HCO_3^-) varied considerably with time.

Concentrations of nitrate and phosphate (given as PO_4 and NO_3, not $PO_4.P$ and $NO_3.N$) are tabulated by Al-Sahaf as averages from analyses during 1967–1969 for the Tigris, Euphrates, Diyala River, and Shatt al-Arab. As *all* values for PO_4 are given as 0.1 mg/l, no confidence can be placed in their accuracy. The values for nitrate vary between 0.2 and 1.0 mg NO_3/l in the Tigris, 0.3 and 1.0 mg/l in the Euphrates, 0.4 and 2.0 mg/l in the Diyala River, and 0.4 and 2.2 mg/l in the Shatt al-Arab. Without further information on the analytical methods, and seasonal variability, interpretation is difficult. In other rivers, such as the Nile (Talling 1976), nitrate may reach notably high concentrations in flood-water, and both it and phosphate can be depleted during phases of phytoplankton growth. These features cannot be elucidated with certainty from the published information on Mesopotamian waters. Al-Hamed (1966, Table 7) lists monthly concentrations of nitrate and phosphate in the Tigris at Baghdad during 1958. The values for nitrate are consistently low, below 0.08 mg NO_3/l: those for phosphate vary between 0.12 and 0.875 mg PO_4/l, with the highest values in winter (January–February). These values do not agree with those of Al-Sahaf (1976), nor with those in the much fuller investigation of Saad & Antoine (1978b). In the latter, analyses of nitrate, nitrite, phosphate, and silica were made from several depths at 12 stations along the Tigris between Baghdad and Qurna (*Fig. 31*). Concentrations of nitrate were especially high (>400 μg NO_3/l) in October, and there was a pronounced fall between April and July in which the end of flood water, uptake by increasing phytoplankton, and bacterial denitrification at higher temperatures may be involved. By contrast, phosphate was particularly high in July, often exceeding 1000 μg PO_4/l, although accompanied by denser phytoplankton. Concentrations of soluble silicon (given as SiO_2) varied rather little, between 7.70 and 9.84 mg SiO_2/l. Thus there appear to be large amounts of P and Si, but not of N, for the strong growth of phytoplankton in July. No estimations of total phosphorus content appear to be published for any Mesopotamian fresh water.

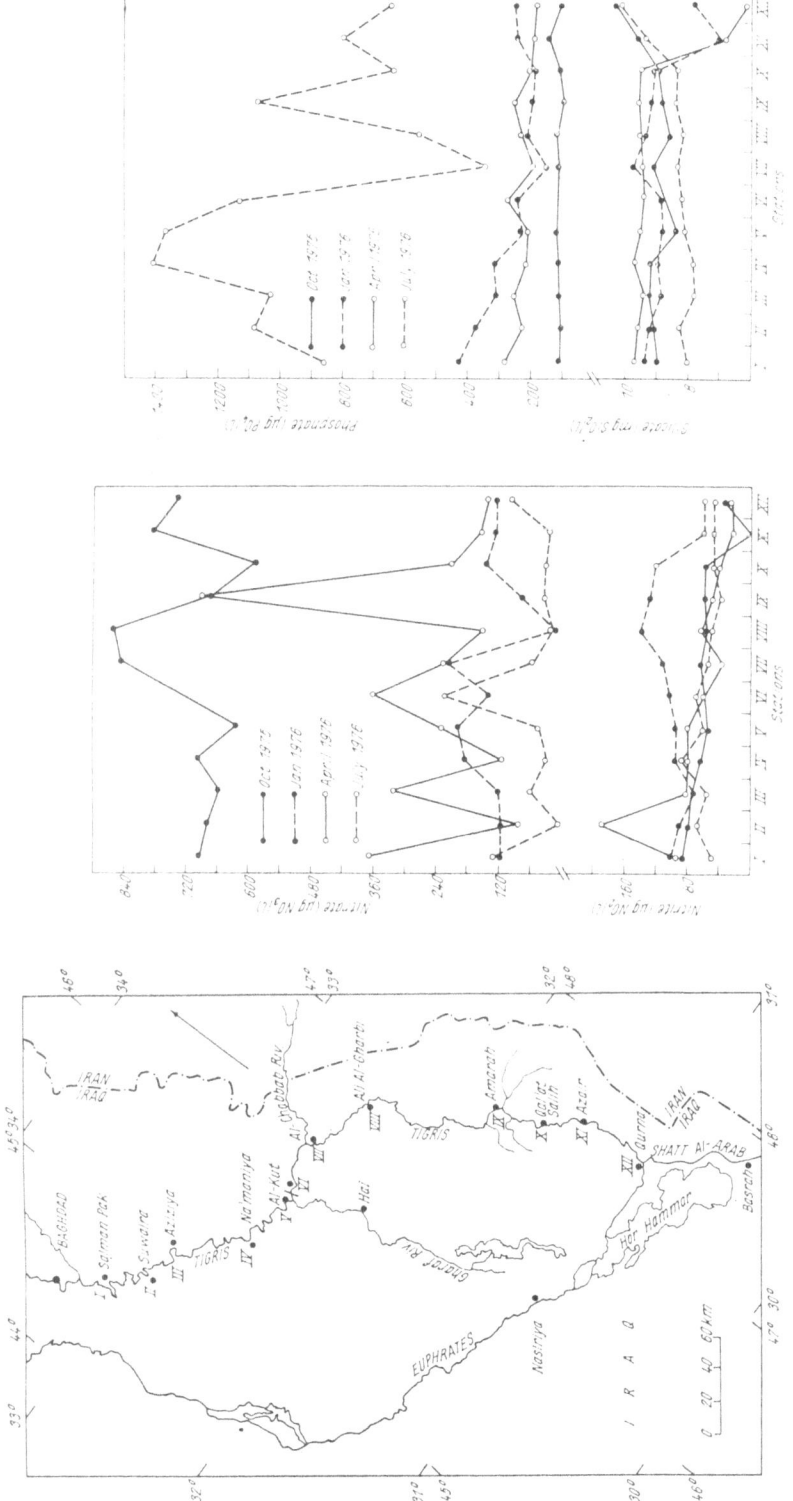

Fig. 31. Concentrations of nitrate, nitrite, phosphate and silicate (expressed as silica) along stations I–XII of the lower Tigris during four months in 1975–6. Concentrations are average values from several depths. (From Saad & Antoine, 1978b.)

77

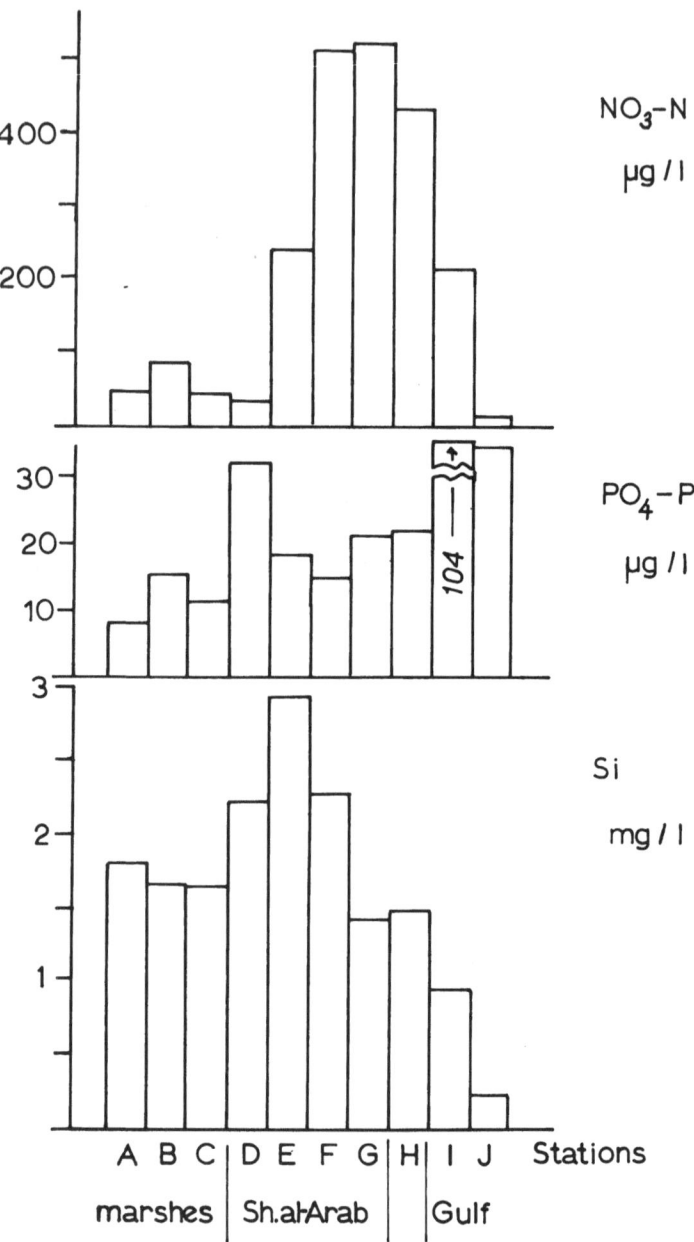

Fig. 32. Distribution of concentrations of nitrate-nitrogen, phosphate-phosphorus and silicate-silicon along a downstream section from marshes near lake Hammar-Shatt al-Arab–estuary in February 1978. Average values from several depths. (From Maulood *et al.* 1979.)

Some recent information is available for the Shatt al-Arab and three adjacent marshes, from sampling in February 1978 (Maulood *et al.* 1979). It is summarized in *Fig. 32*, as average values (from several depths) along a longitudinal section from marshes to Gulf. The conspicuous decline in Si from fresh to sea water is to be expected, being commonplace in estuaries. Although some influence of marine diatom growth may have occurred, the authors considered that such growth was unlikely to have been limited by the lower silicon concentrations. The large decreases in nitrate (and total inorganic) – nitrogen were, they believed, more suggestive for growth limitation, particularly as the concentrations of phosphate-phosphorus were relatively considerable. Information on seasonal changes is clearly required. If the relatively low concentrations of inorganic ($NO_3^- + NO_2^- + NH_4^+$) nitrogen, and fairly low N/P ratios, found in the marsh waters are generally representative, they may have considerable implications for the ecology of this large region. A combination of low inorganic nitrogen (0.02 to 0.05 mg NO_3/l) and considerable inorganic phosphorus (0.78 to 0.9 mg PO_4/l) is also briefly indicated by Al-Kaisi (1964) for another standing water, Lake Habbaniyah. Thus, from the scanty published information available for all major types of Mesopotamian waters, it would be dangerous – as for African freshwaters – to follow the common assumption that phosphate is the most prevalent limiting nutrient.

I am greatly indebted to Dr. J. Rzóska for much help, encouragement, and material used in compiling this and the following section.

6a Phytoplankton

J. F. Talling

By analogy with other large tropical and sub-tropical river systems, such as the Nile, a development of phytoplankton could be expected in the Tigris-Euphrates. Its reality was indicated by casual observations of Al-Hamed (1966), who mentioned (without details of localities or species) a diatom-dominated phytoplankton derived mainly from the genera *Synedra*, *Tabellaria*, *Melosira*, *Cyclotella* and *Fragilaria*.

Later observations on a single day, 17 April 1974, were carried out by Saad & Kell (1975) for the two rivers just before their ('old') junction at Qurna, as well as for the Shatt al-Arab downstream. They list 138 species, including many diatoms. Quantitative estimations showed that the diatom *Synedra ulna* was strongly dominant (214–324 cells/ml) in the Tigris and Euphrates stations, followed by *Diatoma elongatum*, *Bacillaria paxillifer*, *Synedra tabulata* (Euphrates), *Gomphonema acuminatum* and *Cocconeis pediculus* (Tigris), *Navicula halophila*, *Nitzschia spectabilis*, and the green algae *Pediastrum boryanum* and *Scenedesmus acuminatus*. Differences of concentration with depth were small. Although there was a general similarity between the two rivers, some species were only found in one (e.g. *Gomphonema acuminatum* and *Cocconeis pediculus* in the Tigris; *Surirella ovata*, *Synedra tabulata*, and *Volvox africanus* in the Euphrates). An intermingling occurred in the Shatt al-Arab downstream. A striking feature is the abundance of species, such as *Synedra ulna*, *Cocconeis pediculus*, and *Gomphonema acuminatum*, more typical of attached (periphyton) communities, from which their upstream 'inocula' may have been recruited.

A more extensive study in time (October 1975–July 1976) and space along the lower Tigris (between Baghdad and Qurna) was made by Saad & Antoine (1978c). However, only generic identifications are given; 47 genera were represented. Diatoms were again strongly represented, especially of the genera *Cyclotella*, *Navicula*, *Nitzschia* and *Synedra*. There were also considerable numbers of green algae, especially *Chlamydomonas*, *Crucigenia*, *Pandorina*, *Pediastrum*, and *Scenedesmus*, and small quantities of blue-green algae. The quantitative estimations in 3 months, January, April and July 1976, showed a strong maximum in July; this may well be related to the subsidence of flood-water. In July, but not the other months, algal densities decline from near Baghdad southwards (*Fig. 33*). Differences with depth were generally inconsiderable.

An almost consecutive study (September 1976 to May 1977) of the phytoplankton of the Shatt al-Arab near Basrah was made by Huq *et al.* (1978) (cf. also Al-Saadi *et al.* 1979). Their counts indicated some decline of total density in mid-winter (December) as well as during flood water in April. Diatoms were again predominant, and the more abundant included such

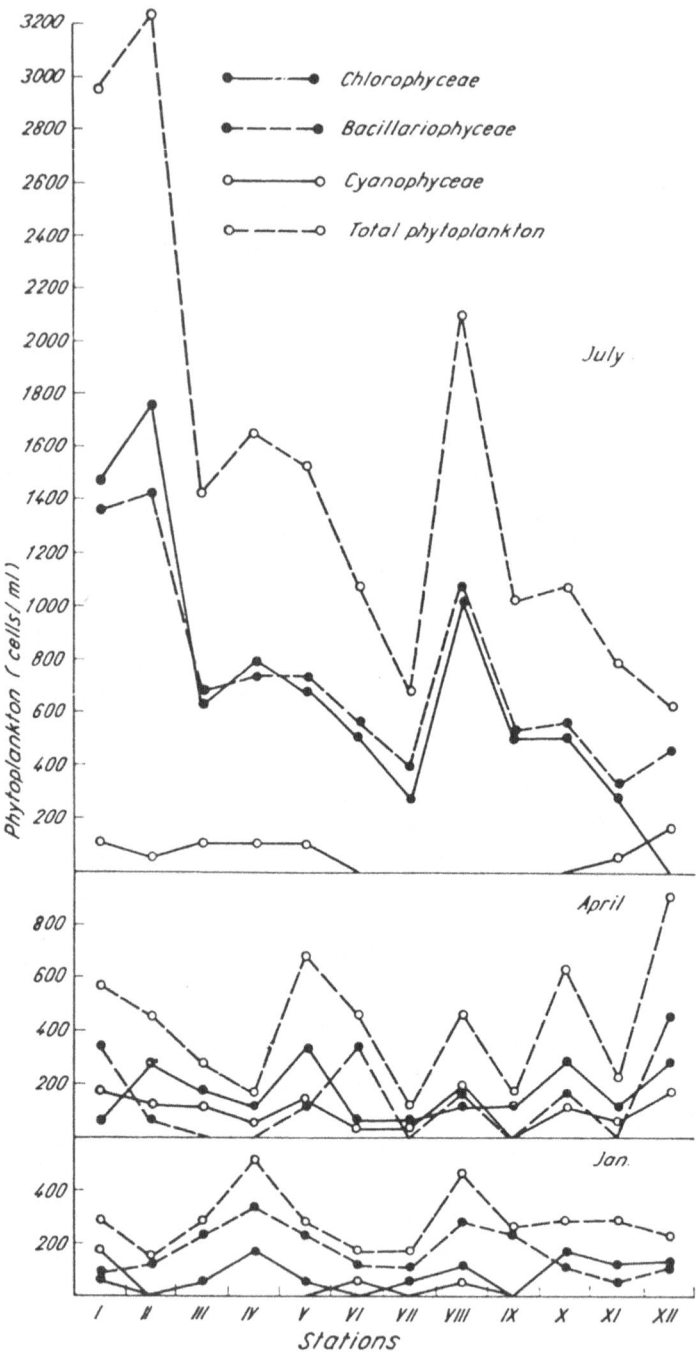

Fig. 33. Seasonal variations in the abundance of *Chlorophyceae,* diatoms (*Bacillariophyceae*) blue-green algae (*Cyanophyceae*) and the total phytoplankton of surface waters of the lower Tigris between Baghdad and Qurna during 1976; for location of stations, see fig. 31. (From Saad & Antoine 1978c.)

typically benthic or periphyton species as *Cocconeis placentula, Rhoicosphenia curvata,* and *Synedra ulna.* In the Shatt al-Arab one would expect a meeting of marine and freshwater elements, with some correlations to the longitudinal gradient in salinity and chloride content. This is confirmed by the studies of Kell & Saad (1975) in April 1974, and of Maulood *et al.* (1979) and Hinton & Maulood (1979) in February 1978. The former list 226 species, including 95 of marine origin whose proportions increased downstream (*Fig. 34*). Diatoms were again generally preponderant; their silica remains contribute to the sediments (Saad 1975). Maulood *et al.* found a large increase of phytoplankton density at their seaward station (*Fig. 28*).

Maulood *et al.* (1979) and Hinton & Maulood (1979) have also described the phytoplankton at three positions in the freshwater marshes north of Lake (Hor) Hammar. They found a considerable diversity of species, with diatoms dominant, and including some considered indicators of brackish water, alkaline water, or hard water conditions. A number occur that have already been noted from the lower Euphrates and Tigris, often more characteristic of benthic habitats. Elsewhere, a secondary *tychoplankton* composed of dislodged benthic diatoms has been described by Maulood & Hinton (1979) from the Serchinar stenothermal spring.

The ecology of phytoplankton in a plain-depression basin is known only from the seasonal and partly quantitative study of Al-Kaisi (1964) on Lake Habbaniyah and adjacent waters. He found a varied community which

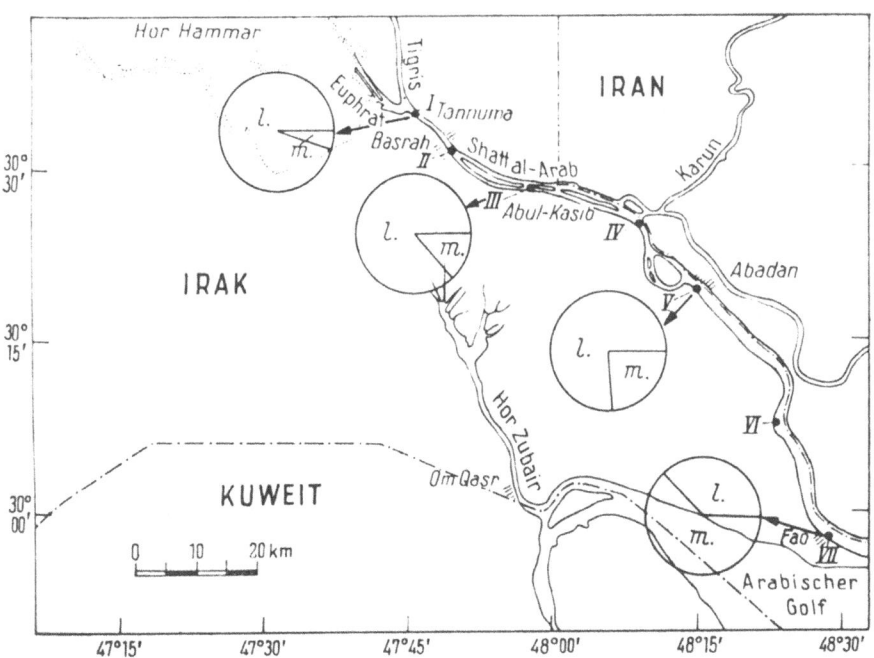

Fig. 34. Percentage proportion of marine (m) and freshwater (l) phytoplankters from stations I – VII along the Shatt al-Arab in April 1974. (From Kell and Saad, 1975.)

83

included diatoms, green and blue-green algae (e.g. *Microcystis aeruginosa*); species of the last group (e.g. *Chroococcus minutus*, *Spirulina subsalsa*, *Synechocystis* sp.) were dominant in the more saline and adjacent lake of Abu-Dibbis. Especially notable in Lake Habbaniyah was the occurrence, occasionally abundantly and at high temperatures (e.g. 28°C), of the cosmopolitan dinoflagellate *Ceratium hirundinella*. This species, with others, showed a definite seasonal periodicity, and distinctive patterns of vertical distribution which involved sub-surface maxima. A number of species showed a small maximum in June, after the clearance of unfavourable turbidity due to flood-water in March–April. Benthic algae, especially diatoms, were also well developed. As in the southern waters, a number of species indicative of saline conditions were noted.

Very little is known about the phytoplankton of the deeper northern lakes. That of Dokan reservoir was studied in 1972–1973 by Al-Hamed (1976); he records a dominance of dinoflagellates, *Ceratium* and *Glenodinium* spp., with maxima in October and April. The outflow river of Derbendy-Chan (Derbendikhan) was found by Hinton & Maulood (1979) to contain the planktonic diatoms *Cyclotella comta* var. *oligactis* and *C. kutzingiana* var. *radiosa*. From the earliest extensive collections of freshwater algae made in the region, by Handel-Mazzetti in 1910, Kolbe & Krieger (1942) described large numbers of the brackish-water diatom *Chaetoceros wighamii* in the plankton of Chattaniye Lake (=Khaturniya, west of Mosul (Syria)). They commented – in advance of chemical analyses – on the prevalence of brackish or halophilic species of diatoms in the lowland plains, and of calciphilic species in the Kurdistan mountain region.

In this limited number of surveys, the minor role played by species of the diatom genus *Melosira* is remarkable, considering their importance in many other river systems – including the Nile. In Mesopotamia, species of *Cyclotella* (surveyed by Al-Kaisi 1974) would appear as more characteristic among the few typically planktonic algae of the river-system. However, occasional abundance of the diatoms *Synedra ulna* and *Cyclotella* spp. in downstream areas has also been observed in the plankton of the lower Nile at Cairo (Belcher, Swale & Talling, unpublished). The relative scarcity of blue-green algae in the plankton of the Tigris – Euphrates may well be connected with lesser development of reservoir storage-basins on these main rivers.

References to Chapters 6 and 6a

Al-Hamed, M. I. 1966. Limnological studies on the inland waters of Iraq. Bull. Iraq. Nat. Hist. Mus. *3*, 1–22.
Al-Hamed, M. 1976. Limnological investigations on Dokan reservoir. Bull. Nat. Hist. Centre, Baghdad 7, 91–109.
Al-Kaisi, K. A. 1964. Studies on the algae of a water system in Iraq. Ph.D. thesis, Univ. Coll. N. Wales. Bangor.
Al-Kaisi, K. A. 1974. The genus *Cyclotella* Kütz. from some aquatic habitats in Iraq. Bull. Coll. Sci., Univ. Baghdad 15, 21–40.
Al-Saadi, H. A. & Arndt, E. A. 1974. Some investigations about the hydrographical situation in

the lower reaches of the Shatt al-Arab and the Arabian Gulf. Wiss. Zeitschr. Univ. Rostock, Math.-Nat. Reihe 23, 517–523.

Al-Saadi, H. A., Arndt, E. A. & Hussain, N. A. 1975. A preliminary report on the basic hydrographical data in the Shatt al-Arab estuary and the Arabian Gulf. Wiss. Zeitschr. Univ. Rostock Math.-Nat. Reihe, 24, 797–802.

Al-Saadi, H.A., Pankow, H. & Huq, M.F. (1979). Algological investigations in the polluted Ashar Canal and Shatt al-Arab in Basrah (Iraq). Int. Rev. Hydrobiol. Hydrogr. 64 (4) (in press).

Al-Sahaf, M. 1975. Chemical composition of the water resources of Iraq. (Russ.) Vodn. Resur. [1975] (4), 173–185.

Arndt, E. A. & Al-Saadi, H. A. 1975. Some hydrological characteristics of the Shatt al-Arab and adjacent areas. Wiss. Zeitschr. Univ. Rostock, Math. Nat. Reihe 24, 789–796.

Brinkmann, W. L. F. 1967. Grundlagen der Bewässerungswirtschaft am mittleren Euphrates. Zeitschr. Bewässerungswirtsch. 1, 65–77.

Hadi, R. & Al-Saadi, A. 1977. Preliminary studies on some major nutrients in the north west Arab Gulf. The Arab Gulf 8, 23–29 (cited from Maulood et al. 1979).

Hinton, G. C. F. & Maulood, B. K. 1979a. Some diatoms from brackish water habitats in southern Iraq. Nova Hedwigia 31 (in press).

Hinton, G. C. F. & Maulood, B. K. 1979b. Freshwater diatoms from Sulaimaniyah, Iraq. Nova Hedwigia 31 : 449–466.

Huq, M. F., Al-Saadi, H. A. & Hamed, H. A. 1978. Phytoplankton ccology of Shatt al-Arab River at Basrah, Iraq. Verh int. Verein. theor. angew. Limnol. 20, 1552–1556.

Kell, V. & Saad, M. A. H. 1975. Untersuchungen über das Phytoplankton und einige Umweltparameter des Shatt al-Arab (Irak). Int. Revue ges. Hydrobiol. Hydrogr. 60, 409–421.

Kolbe, R. W. & Krieger, W. 1942. Süsswasseralgen aus Mesopotamien und Kurdistan. Ber. dtsch. bot. Ges. 60, 336–355.

Mahmoud, T. A. & Sh. Ahmad. 1979. A water quality study of a stretch of the river Tigris. Water Research 13: 785–790.

Maulood, B. K. & Hinton, G. C. F. 1978. An ecological study on Serchinar Water – chemical and physical aspects. Zanco, Sci. J. Univ. Sulaimaniyah 4 (3), 93–117.

Maulood, B. K. & Hinton, G. C. F. 1979. Tychoplanktonic diatoms from a stenothermal spring in Iraqi Kurdistan. Br. Phyc. J. 14, 175–183.

Maulood, B. K., Hinton, G. C. F., Kamees, H. S., Saleh, F. A. K., Shaban, A. A. & Al-Shahwani, S. M. H. 1979. An ecological study of some aquatic ecosystems in southern Iraq. Tropical Ecology 20 (1).

Mohammad, M-B. M. 1965a. Preliminary observations on some chemico-physical features of the Shatt al-Arab estuary. Proc. Iraqi Sci. Soc. 6, 34–40.

Mohammad, M. B. M. 1965b. Further observations on some environmental conditions of Shatt al-Arab. Bull. Biol. Res. Centre, Baghdad 1, 71–79.

Saad, M. A. H. 1975. Distribution of diatom-silica in the sediments of Shatt al-Arab and Arabian Gulf. J. Arab Gulf Univ. Basrah Iraq 3, 199–211 (cited from Saad & Antoine 1978b).

Saad, M. A. H. 1977. Studies on the bottom deposits of the lower reaches of Tigris and Euphrates, Iraq. Bull. Coll. Sci. Univ. Basrah 6 (cited from Saad & Antoine 1978a).

Saad, M. A. H. 1978a. Some limnological studies on the lower reaches of Tigris and Euphrates, Iraq. Acta Hydrochimica et Hydrobiologica 6, 529–539.

Saad, M. A. H. 1978b. Seasonal variations of some physico-chemical conditions of Shatt al-Arab Estuary, Iraq. Estuarine & Coastal Marine Science 6, 503–513.

Saad, M. A. H. & Antoine, S. E. 1978a. Limnological studies on the River Tigris, Iraq. 1. Environmental characteristics. Int. Revue ges. Hydrobiol. Hydrogr. 63, 685–704.

Saad, M. A. H. & Antoine, S. E. 1978b. Limnological studies on the River Tigris, Iraq. II. Seasonal variations of nutrients. Int. Revue ges. Hydrobiol. Hydrogr. 63, 705–719.

Saad, M. A. H. & Antoine, S. E. 1978c. Limnological studies on the River Tigris, Iraq. III. Phytoplankton. Int. Rev. ges. Hydrobiol. Hydrogr. 63, 801–814.

Saad, M. A. H. & Antoine, S. E. 1979a. Seasonal distribution of dissolved organic matter in the River Tigris, Iraq. Limnologia (in press) (cited from Saad & Antoine 1978c).

Saad, M. A. H. & Antoine, S. E. 1979b. Data on transparency, suspended matter, total residue, fixed total residue and volatile matter in Tigris, Iraq. Bull. Coll. Sci. Univ. Basrah (in press) (cited from Saad & Antoine 1978c).

Saad, M. A. H. & Kell, V. 1975. Observations on some environmental conditions as well as phytoplankton blooms in the lower reaches of Tigris and Euphrates. Wiss. Zeitschr. Univ. Rostock *24*, 781–787.

Talling, J. F. 1976. Water characteristics. In: *The Nile, biology of an ancient river* (ed. J. Rzóska), pp. 357–384. The Hague. Junk.

Talling, J. F. & Talling, I. B. 1965. The chemical composition of African lake waters. Int. Revue ges. Hydrobiol. Hydrogr. *50*, 421–463.

7 General Biology of Iraq waters

An additional remark must be made to the above chapter on Phytoplankton of Iraq. In Chapter 4 attention was drawn to the wider panorama of waters in countries adjacent to present Iraq. Some features like salinity, instability of levels and the impact of desert conditions are common to a wide area of the Near East. In fact Syria and parts of southern Turkey would be included into an overall limnology, but this is a task beyond the scope of this book. The discovery of *Chaetoceros wighamii,* a saline form in lake Chattaniye (Khatuniya) in Syria, as the dominant phytoplankton form indicates the wide spread of saline conditions in waters worthy of investigation (Handel-Mazzetti 1914; Kolbe & Krieger 1942). Another field requiring attention is the limnology of temporary waters mentioned for the Azraq oasis in Chapter 4 and alluded to in remarks below for Iraq.

Water vegetation in Iraq waters

Astonishingly there seems to be a dearth of studies on water vegetation in the available sources. Handel-Mazzetti has described a.o. the riverine vegetation as result of his travels (1914), but only few descriptions have been made and these not by botanists but by nature lovers and enthusiastic travellers. The books on the 'Marsh Arabs' by Thesiger (1964) and Young (1977), contain a veritable natural history of the great southern marshes.

The marshes extend over an area of 6000 miles^2 or according to another estimate to 15 000 km^2. An archipelago of islets of firmer grounds is surrounded by vast expanses of reeds, *Phragmites* and *Typha* being the principal components. These may reach a height of 8 m and are important for the economy of man and the animal world. These marshes must be old as testified historically and by the perfect adaptation of the Arab tribes living there; apparently Hor Hammar the largest water body, originated or extended in size in the 6th century A.D.

The books by Thesiger and the later one by Young are superbly illustrated, and provide an insight into the ecology of this unique area. A photograph supplied by the Iraq Cultural Centre in London shows some essential features (*Fig. 35*). *Phragmites* is used as building material for dwellings as there is no stone; the reed sprout is used as fodder for live stock and as cover for the unstable ground. Great artistry is shown in the rest houses and assembly dwellings, these almost resemble Gothic cathedrals, built of skilfully bound bundles of the reed. To facilitate new and tender growth selected stands are fired and harvested as fodder for water buffaloes, sheep and goats; communication is by boats surprisingly elegant and efficient made of riverine poplar wood. Fishing is done by some tribes, plots of barley, wheat and rice are on higher grounds. The whole economy is or, shall I say, was based on the natural resources in a perfect way. It reminds me of the Nilotic tribes in the

Fig. 35. Marsh Arabs, settlement amidst the islets of firmer grounds in the southern marshes; *Phragmites* is the skilfully used building material. (From Iraq Cultural Centre, London.)

Sudd region of the Sudan. Both habitats are now threatened by drainage projects.

Submerged vegetation is abundant; water lilies *Nymphaea coerulea*, *Nymphoides peltata* and *indica* flourish, besides *Ceratophyllum*, *Myriophyllum*, *Echinochloa* and the South American invader *Salvinia*. Vast areas of the banks are covered by *Polygonum*, with a profusion of annuals.

Animal life in the marshes and waters

Animal life in the marshes impressed visitors; of mammals the wild boar lives in the reed thickets, otters live in the waters, rich in fish; reptiles and amphibians abound but have not been named. Thesiger tells of old memories of lions lingering in peoples' tales and this seems plausible. The most striking group are the birds; the resident water fowl are reinforced each year by migrants from the north, waders, cranes, storks and ducks 'in myriads'. At least part of this unique area should be preserved, and a study of the marshes should be made, in addition to plankton observations.

When talking of hydrobiology the instability of all but especially standing waters should again be mentioned. The Shattra group of four basins between the Tigris at Kut al Imara and Nasyriya on the Euphrates occupy 800 km² during the floods, but only 250 km² at low water and like most of these inundation marsh waters decrease in depth from over 1 m to 0.5 in the low season. The Suweicha basin dwindles from an area of about 1000 km² to only 350 km²; the Saniya, adjacent to the Tigris, varies from 700 km² to half that size. Thus space photos show changes according to the season. Some fringes become dry and these fringes are transitions to temporary waters. There is therefore scope for 'desert Limnology' in Iraq, considering that the larger part of the country is desert. I have mentioned the term and its manifestations in the Jordanian Azraq Oasis in Chapter 4, and this area is confluent with the Iraq desert.

Only one author describes the aquatic fauna characteristic of temporary waters in Iraq; Gurney, who had material from waters around Amara and Baghdad, collected by the desert explorer Buxton and others. In a paper published in an issue of the Bombay Nat. History Society (vol. 27, 1920–1921) he gave a list of the Crustacea found, which is still the most accurate account. This list is given in full. Part of the material came from northern Persia; in Iraq marshes, inundation basins, drainage ditches and temporary pools were sampled; some of the identifications may have been subsequently amended, but this does not change the value of the paper.

Malacostraca:

Sesarma boulengeri Calman taken at Basra in freshwater, though it occurs at Fao in brackish water.

Potamon fluviatile var. *ibericum* M. de Bieberstein taken in Persia near the Caspian but also

Potamon fluviatile var.? at Qalat Saleh on the Tigris.

Phyllopoda:

Artemia salina var. *asiatica* Fisher-Dadyay in saline pools near Amara, no males.

Apus asiaticus nom. nov. and *Apus granarius* Sars from vicinity of Baghdad; (these identifications are now overtaken by the revision by Longhurst 'A review of the Notostraca' Bull. Brit. Mus. N.H. 1955, 3 and the generic name is now *Triops*).

Leptestheria sp. found at Amara; Gurney believed this to be a new form.

Gladocera:

Daphnia lumholtzi Sars, at Amara and in marsh at 'Ezra's Tomb' between Amara and Basra.

Daphnia magna Strauss, in flood waters at Amara, males present but no ephippial females.

Daphnia pulex de Geer forma obtusa, in Persia.

Daphnia longispina O.F.M., at Ezra's Tomb near Tigris and Persia.

Ceriodaphnia reticulata Jur. at 4 sites in Iraq and Persia.

Moina rectirostris Jur. 4 sites in Iraq and Persia.

Moina dubia Richard near Amara.

Simocephalus exspinosus Koch and *S. vetulus* O.F.M. at Amara and in Persia.

Copepoda:

Cyclops vicinus Ulianin widespread in the Palaearctic deserves special mention.

Cyclops viridis, *C. vernalis*, *C. bicuspidatus*, *C. leuckarti*, *C. crassus*, *C. albidus*, *C. agilis*, *C. affinis*, *C. bicolor*, *C. diaphanus*, all widespread, either cosmopolitan or palaearctic. Gurney describes and illustrates a 'new' species *Cyclops buxtoni* n.sp.

Canthocamptus staphylinus Jur. the only Harpacticid found is cosmopolitan and palaearctic.

Of greater interest are the Diaptomids: *Diaptomus vulgaris* Schmeil taken at Ezra's Tomb and in Persia, palaearctic. *Diaptomus blanci* in irrigation canals at Amara, Asiatic species. *Diaptomus chevreuxi* Guerne & Richard, found in North Africa, Egypt; found at Gantra Sarut near Tigris. (Gurney gives excellent drawings). A *Diaptomus* sp. from Persia not determined.

Ostracoda:

Notodromas persica n. sp. is described and illustrated, collected at Resht, *Cyprinotus dentatomarginatus* Baird-Sars, *C. incongruens* Rajd., *Cypris virens* Jur.-Mull., *C. pubera. Mull.*, *Potamocypris variegata* (Brady-Norman), *Herpetocypris reptans* (Baird), *Ilyocypris bradyi* Sars. It seems that the Ostracods found are palaearctic.

It is possible that one of the two crabs mentioned by Gurney is identical with *Potamon potamios*, which is widespread in the Near East and has been found in the food debris of prehistoric man as mentioned in Chapter 3. On the whole only some species in Gurney's list are of zoogeographical interest. Most are palaearctic and widespread. Ecologically interesting are the Euphyllopods, as already mentioned.

Subsequent papers are lists of collections of Crustacea without ecological background. Al Hamed (1966) mentioned a 'fairy shrimp', possibly *Artemia salina.* Mohammad (1965) has published a list of Cladocera collected in 10 areas of Iraq; though valuable as a start, it does contain mainly palaearctic ubiquists, except *Podon polyphemoides* (Leuckart) a brackish form found in the delta region.

A well documented list of 'Littoral Cladocera of Iraq' was given by Khalaf and Smirnov (1976) from riverine and standing water shores; the thorough identifications are by Smirnov, an authority on the chydorids; 16 species of Chydorids, 5 of *Macrothricidae* and 2 daphnids were found. Most of the species found are commonplace but 12 are in Smirnov's opinion of interest and belong to the 'southern geographic range'. In my opinion some are circum-tropical and have been found in the Nile valley. Nevertheless this shows the mingling of palaearctic and warm water elements including Asian influences.

Only one paper deals briefly with true zooplankton, Al Hamed (1976) gives some few data on densities of copepod and rotiferan plankton in the Dokan reservoir.

Benthos

The same paper by Al Hamed contains information on the Dokan bottom fauna. *Limnodrilus hoffmeisteri,* determined by Brinkhurst, forms 76 per cent of the benthos by numbers, 17 per cent by weight, *Planorbidae* and *Chironomus plumosus* and *Ch. digitalis* are present, 15 species of fishes have populated the reservoir. There is no littoral vegetation due to strong level changes.

In the above survey of available sources on the animal world of Iraqi waters fishes have been omitted; existing knowledge is scanty and not critical. Taxonomic problems are difficult in this area and only an institution with large comparative and bibliographic resources can elucidate the problems. Dr. Banister of the British Museum of Natural History has responded to my request and his chapter is the first critical examination of the Mesopotamian–Iraq fish fauna.* Similarly Dr. Talling, with his rich experience of warm water limnology, has written the chapter on water characteristics and phytoplankton on the basis of existing sources.

These are meagre and restricted; no specific studies have been made on the rivers. Admittedly work on long rivers is difficult and only recently have modern studies on the biological regime of rivers appeared (Danube, Liepolt 1967, Nile 1976, Volga 1978) after some good, earlier starts 50 years ago.

In my opinion concentrated studies on the bottom fauna are now a priority with the help of accurate taxonomic determinations, and in connection with

* Boxhall (1976) has described a new genus and two new species of copepod parasites from fishes in the Tigris at Mosul; one of the new species is *Pseudolamproglena annulata* from *Cyprinion macrostomus.*

this the food of the fishes should be examined. Al-Hamed (1966) has mentioned the importance of fisheries in the waters of Iraq. Not a single study has so far appeared on the zooplankton of the rivers, in which the truly planktonic forms were separated from the adventitious species; the seasonal appearance and other problems of quantitative, longitudinal distribution have not been touched. A wide scope for fruitful scientific and practical biological work exists.

Anopheles and Malaria in the Near East

Under this title a book published in 1950 by the London School of Hygiene and Tropical Medicine, contains three studies on malarial vectors in Syria and Lebanon, Trans-Jordan and Palestine, Iraq and Persia. All these studies were made during the Second World War (1939–1944) as part of protection measures by the British army.

In all these arid regions water-born mosquito vectors can breed only where irrigation installations exist which cause pools and spillage. In the first paper Leeson did not find mosquito larvae in the Syrian Euphrates and only one adult near the river from Heskeneh to Abu Kemal, a significant ecological feature. Pools created for watering livestock were infested in many places by *Anopheles superpictus*.

Lumsden and Yofe found malaria endemic in Palestine and Jordan, depending on the presence of standing waters. Hence in large parts of the area mosquitoes can only breed after the seasonal winter rains, which form mostly temporary waters lasting for a short time, up to 5 weeks. The Azraq Oasis contains spring-fed permanent pools and, after the rains, large areas of gradually drying out 'playas' and small waters in some of the desert wadis. With rapid evaporation, salinity increases so as to make the water toxic later except for some animals, as described in Chapter 4. Malaria is here endemic and the list of mosquito species shows a predominance of Mediterranean and palaearctic forms.

Iraq and Persia were traversed in most parts by Macan. He goes as far as to say that malaria in Iraq is largely 'man-made', vectors breeding in neglected channels, ditches, pools left by inundations and quiet river bays, overgrown by vegetation (*Fig. 36*). The distribution of mosquito elements is governed by the climate, oriental species prevail in the alluvial basin, palaearctic forms in the foothills. Oriental species are unable to colonize areas with winter frosts, whereas palaearctic forms probably cannot survive the combination of high temperatures and low humidities in the plain. The 'oriental' element extends into Arabia; from there westwards African species appear in Egypt thrusting north via the Nile valley as shown in the Nile monograph (1976). Palaearctic forms include Mediterranean influences. Some traits of mosquito distribution recall other biogeographic zones, e.g. vegetation.

The most important malaria vector in the plain seems to be *Anopheles stephensi* whose distribution coincides with irrigation installations, where vigilance has been relaxed.

Fig. 36. River Bay at Hilla on the Euphrates, breeding ground for aquatic life including mosquitoes. (Courtesy Iraq Petroleum Co., London.)

A list of species collected is given on page 146 of the book, divided into two geographic areas and further subdivided into four categories (a, b, c, d) according to importance.

	Oriental		Palaearctic
a	Anopheles stephensi	*a*	Anopheles saharovi
			A. maculipennis
			A. superpictus
b	A. pulcherimus	*b*	none
c	A. hyrcanus	*c*	A. claviger
	A. fluviatilis		A. dthali
			A. marteri
			A. algeriensis
			A. multicolor
d	A. culicifacies	*d*	A. sogdianus
	A. turkhudi		A. sergenti
	A. arabicus		

Category *a* denotes an important vector; *b* contains widespread forms, numerous but with restricted distribution; *c* rare and in small numbers; *d* are forms of little significance. The nomenclature is based on the 1943 work by Smart.

Macan gives information on occurrence, breeding habits and localities for each species. Malaria in Iraq, according to Macan, is intense in places but sporadic and unpredictable because it depends on local irrigation maintenance. The rivers become breeding places when they spill; channels and side bays especially of the Shatt al-Arab were highly malarious during the time of investigation.

I owe Dr. T. T. Macan great gratitude for all material on which the above account is based.

8 The fishes of the Tigris and Euphrates rivers

K. E. Banister

The inhabitants of the Fertile Crescent have been involved with the fishes of the Tigris and Euphrates rivers for over 4000 years. The oldest-known fishponds were built in temples by the Sumerians and, before long, many of the settlements throughout the region had their own fishponds. In view of the long-established interest in fishes in this part of the world, it comes as a surprise to realize that our present-day knowledge of the fish fauna and its zoogeographical affinities is extremely scanty.

The headwaters of the Tigris and Euphrates originate in the highlands of Turkey, and are separated by high, narrow watersheds from rivers flowing into the Black and Caspian Seas. To the northwest, much lower watersheds in the arid areas of Syria and southern Turkey separate the Euphrates from rivers draining into the Mediterranean. The lower reaches of the Tigris and Euphrates are confined between Iran's Zagros mountains in the east and the Alwadyan desert of Iraq and Arabia to the west.

Below Baghdad, the rivers flow through a marshy, alluvial plain until they join at Basrah to form the Shatt-el-Arab, which shortly empties into the Persian Gulf. Interconnections between the Tigris and Euphrates are plentiful and there is no doubt that the two rivers can be treated as a single unit from a zoogeographical point of view. The headwaters arise close together to the east of Lake Van, whilst at Basrah the rivers join. Elsewhere in the system there are small, natural interconnections as well as many artificial irrigation channels.

There are some 'lakes' (standing waters) within the catchment area. The isolated Lake Van basin, although formerly part of the system and possessing certain faunal similarities (Kuru, 1971), is not considered in this chapter. In the south there are extensive swamps with areas of open water. The largest of these is Hawr al Hammar with a surface area of 5200 km² at high water and 3500 km² at low water level. The immense volume of silt deposited by the Tigris and Euphrates has made Hawr al Hammar shallow and eutrophic, and it is a major centre for freshwater fisheries. Mileth Thartar is a shallow sump fed by the Wath Thartar which lies between the Tigris and Euphrates and is connected to the latter by a canal. Mileth Thartar is extremely variable in its dimensions. The deepest lake in the system is Habaniyah, to the west of the Euphrates, which is only 13 m deep when flooded and 6 m deep at the end of summer. The swamps, which can be regarded as eutrophic lakes, more than double in area when they are flooded, at which time local basins may be 3 m deep.

The present-day disposition of water in the Tigris and Euphrates catchment is ephemeral. Archaeological research has provided ample evidence for floods of both salt and fresh water at various sites around the lower reaches of the rivers. The present-day catchment, the Mesopotamian Plain, came into existence at the end of the Zagros orogeny (Miocene-Pliocene) and has been gradually subsiding ever since (Lees & Falcon, 1952). These authors suggested that prior to the Zagros orogeny the rivers from central Iran flowed along mature valleys to the Tigris and Euphrates and/or the Persian Gulf. (It should be noted that their opinion has not been substantiated). Later uplift caused increased erosion producing the river terraces or isolated plateaux. The apparent upstream movement of the delta in the immediate post-glacial period was caused more by a rise in sea-level than by the sinking of the land. Lees & Falcon (*op. cit.*) point out that even within the last 2500 years, salt water floods have extended upstream of the junction of the Tigris, Euphrates and Karun Rivers. They argue that these salt water encroachments probably resulted from local subsidence. Indeed, the extensive marshes owe their existence to this phenomenon. The largest marsh, Hawr al Hammar, has a historically recent origin. Lees & Falcon (1952) quote Le Strange (1905) who translated old writings recording the breeching of dykes at the end of the fifth century A.D. and finally, the creation of the great swamp by a massive flood and subsidence in about 636 A.D. (see other interpretations in Chapter 5).

It is axiomatic that, prior to discussions on the zoogeographical affinities of the Tigris and Euphrates ichthyofauna, the composition of that ichthyofauna must be known. This is a problem to which the only solution presently available, and an unsatisfactory one at that, must be based on a review of the literature. There are certain dangers inherent in this approach. Firstly, confirmation of the identification is generally impossible although, in some cases, the alleged presence of a species has been confirmed by reference to collections in the British Museum (Natural History). Secondly, the taxonomic status of some species is in doubt. In certain cases, doubt arises over whether a taxon recognized from the Tigris and Euphrates could have specific or subspecific rank. The difficulties associated with ranking at this level are exaggerated by our inadequate knowledge of infraspecific variation, as Kuru (1971) has shown. Further doubt, due entirely to a lack of detailed study, concerns the generic placement of certain species. Errors already present in the literature have been progressively compounded by subsequent authors who have not checked their original sources. Some of the most spectacular errors are itemized below.

Khalaf (1962) and Mahdi (1962) independently produced the first two check lists in recent times. Both authors list 41 primary freshwater fish species but these two lists have only 27 nominal species in common, thereby implying a total of 55 species in the Tigris and Euphrates Rivers. Al-Hamed (1966) produced a check list essentially the same as Khalaf's (*op. cit.*). Mahdi & Georg (1969) listed 45 freshwater fishes which excluded, without comment, some of the species included in the senior author's 1962 list. The most recent, and most extensive, check list (Al-Nasiri & Hoda, 1976) is a compilation from those of Mahdi & Georg (*op. cit.*) and Khalaf (*op. cit.*), with the addition of a few Syrian species, which the authors thought might live in Iraq, taken from Beckman (1962).

The uncritical use and compilation of check lists can result in some alarming errors. For example, Al-Nasiri & Hoda (1976) include the species *Euglyptosternum coum* (L) in the family Siluridae, *Glyptothorax cous* (L) in the Sisoridae and *Arius cous* Heckel (*sic*) in the Ariidae. In reality, all three names refer to a sisorid catfish. Its inclusion as a silurid stems from a mis-reading of Khalaf (1962). Further confusion originates from Ladiges (1964) who reported the occurrence of *Arius cous* Heckel 1843 (*sic*) in the Tigris. Reference to Heckel, 1843: 1094 shows that this 'ariid' is the sisorid mentioned above and that Heckel (*loc. cit.*) merely referred this species to the genus *Arius* and did not, as Ladiges implies, describe it. Al-Nasiri & Hoda (1976) record *Arius cous* and even provide an illustration of it (Al-Nasiri & Hoda 1976: fig. 101). The descriptions of *Euglyptosternum coum* and *Glyptothorax cous* in Al-Nasiri & Hoda (1976) refer to figures of these two 'species', respectively figs. 67 and 71. Perhaps significantly, there is no fig. 67!

Although it can be established that only one species is involved (the sisorid), even the correct generic name for this species is in doubt. If it is not referable to *Glyptothorax* as limited by Hora (1923) then it must be included in *Aclyptosternon* of Bleeker, 1863). My inclusion of the species *coum* in *Aclyptosternon* is no more than an act of convenience.

The case of *Aclyptosternon coum* is not unique. Species attributed variously to the genera *Leuciscus*, *Alburnus*, *Acanthobrama* and *Abramis* in particular, have appeared, disappeared and been re-assigned in various check lists. However, it must be admitted that, with the recent exception of *Acanthobrama* (Goren, Fishelson and Trewavas, 1973) these genera are extremely ill-defined and are very probably artificial, polyphyletic assemblages.

A particular case in another cyprinid assemblage can be cited. *Systomus albus* Heckel 1843 was synonymized with *Barbus luteus* (Heckel, 1843) by Günther in 1868. Al-Nasiri & Hoda (*op. cit.*) placed *albus* in *Barynotus* Günther in 1868 and their check list contains both *Barynotus albus* and *Barbus luteus*.

Bearing in mind all these various limitations, and especially the lack of adequate revisionary studies at the species level, the Table represents as accurate a faunal list as it is possible to compile. It incorporates the latest, generally accepted revisions, and includes all the nominal species of the check lists mentioned above. Specific comments on the non-acceptance of certain generic re-arrangements advocated by the revisors are discussed below. There are several comments and qualifications that must be made concerning the contents of the Table. For example, *Phoxinellus zeregi* is listed as present by Mahdi & Georg (1969) but, according to Beckman (1962) and Karaman (1972), is only found in the rivers to the west of the Tigris and Euphrates. Similarly, *Garra lamta* is confined to northern India in the opinion of Menon (1964). Should these, apparently anomalous, listings have been based on misidentifications, it is impossible to suggest which species may have been misidentified.

Secondly, a few species in the table (e.g. *Hemigrammocapoeta nanus*) are included on the basis of preserved specimens in the collections of the British Museum (Natural History); they have not previously been recorded from the Tigris and Euphrates. The presence of *Heteropneustes fossilis* in these rivers is a recent event. Khalaf (1962) reported that in 1960 large numbers of this species appeared for the first time in the Shatt-el-Arab and subsequently dispersed through the system. *Heteropneustes fossilis* is native to the freshwaters from Pakistan eastwards and it is assumed that a large scale migration led to its establishment in the Tigris and Euphrates.

Thirdly, there is the problem of generic attribution, a matter of no small importance in the discussion of zoogeographical affinities. For example, when Karaman (1971) revised the *Barbus* species of this region he placed *B. sharpeyi* and *B. luteus*, respectively, in his newly erected genera *Mesopotamichthys* and *Carassobarbus*. *Barbus subquincunciatus* he referred to *Bertinus* of Fang 1943 and *Barbus grypus* to *Tor* of Gray 1834. Karaman's generic re-arrangements have not met with universal acceptance. Banarescu (1973) has expressed doubts about the validity of some of Karaman's genera, and the use of *Bertinus*, in particular, has been criticized by Banister & Clarke (in press). This 'genus' is characterized by the presence of only four teeth in the inner pharyngeal row (five is the usual number in *Barbus*) and the teeth are enlarged and molariform. Very similar modifications to the pharyngeal dentition occur in unrelated *Barbus* species (e.g. *Barbus eurystomus* from Lake Malawi) and it has been argued that they represent a response to a molluscan diet (Banister & Clarke, in press). It would seem most unwise to change the generic attribution on such dubious, or at least, unproven characters. It is, however, extremely likely that not only *Barbus*, but also some of the other 'genera' are polyphyletic. Since detailed phylogenetic analyses are lacking, discussions on the faunal affinities have, of necessity, to be based on what may well be non-monophyletic assemblages.

The primary freshwater fish fauna is dominated by members of the Superorder Ostariophysii or, to put it another way, there is a remarkable absence of non-ostariophysan fishes. Apart from the Trout (which although not a primary freshwater fish included in the Table for convenience) the only non-ostariophysan fish is *Mastacembelus mastacembelus* a member of the Superorder Percomorphii.

The almost complete domination of the fauna by ostariophysans is unique in a sub-tropical river system of this size. Although throughout southern Asia generally the majority of fish species in the rivers belong to the Ostariophysii, there is a substantial contribution from species belonging to other groups (*e.g.* Centropomidae, Channidae and anabantoids). In small rivers, especially in arid areas with a sparse fauna (e.g. Arabia, see Banister & Clarke, 1977), the fauna can be entirely ostariophysan. However, in these cases only a few species are involved (8 in the case of the Arabian peninsula, in contrast to the 59 nominal species in the Tigris and Euphrates).

Apart from the Trout, which is confined to the Anatolian headwaters of the system (Kuru, 1971), nothing is known of the distribution of species throughout the Tigris and Euphrates. Other Anatolian species listed by Kuru (*op. cit*) also occur in subsequent check lists, without any comment, so it must be assumed that they are widely distributed throughout these rivers. There may well be seasonal movements of fishes, as Al-Hamed (1966) reported. This author noted that as the water level fell during the summer, the water warmed up and the fishes either migrated upstream or into the deepest parts of 'lakes' and marshes to reach cooler conditions.

Families and genera of freshwater fishes common to Africa and Asia have long exercised the ingenuities of biogeographers. The discontinuous distribution shown by *Clarias*, *Barilius* and the Anabantidae has suggested to some authors a Gondwanic origin for these groups. Others, however, have disagreed and the opposing views, exemplified by *Clarias* distribution, are collated and discussed in Banister & Clarke (1977). None of the three taxa mentioned occurs in the Tigris and Euphrates; *Barilius mesopotamicus* has now been shown to be a *Leucaspius* (Howes, in press).

Mastacembelus and to an even greater extent, *Garra* present a less clear-cut problem. Both genera have representatives in Africa, Asia as well as in the Tigris and Euphrates, but these genera have a relatively continuous distribution from Africa to eastern Asia (Banister & Clarke, 1977).

The distribution of the family Mastacembelidae is interesting. In Africa many species occur in the Zaire basin, but fewer live in north-east Africa. In Asia, the family is widespread, from the Tigris and Euphrates into Pakistan, India, Sri Lanka, Burma, through the Malay Peninsula to Sumatra and Borneo and in Indo-China as well as southern China. Sufi (1956) recognizes 16 species, belonging to two genera, throughout the Asian region. In the Middle East only one species is present, in the Tigris and Euphrates. Although *Mastacembelus mastacembelus* was described from a river near Aleppo in Syria (variously the river Quweik, Kueik or Kowick) it has not been found there during recent surveys (Beckman, 1962). The River Quweik is a closed system, flowing southwards from the highlands of southern Turkey into the Syrian desert. As *Mastacembelus mastacembelus* is not found in the Orontes river, one is led to presume that the Quweik and Euphrates may formerly have been confluent. The watershed between these two rivers is extremely low and their present separation would seem to be due more to contemporary aridity than to the presence of physiographical barriers. The highlands between the Orontes and the Quweik present a more formidable physiographical barrier to the extension of the range of *Mastacembelus*, although the orogeny of this range has not been dated.

Apart from the Quweik River, *Mastacembelus mastacembelus* is not found outside the Tigris and Euphrates system. Within that system it appears to be widespread although it has not been reported from the headwaters. Mastacembelids are absent from Iran but are represented in Baluchistan by *M. armatus*, which also occurs in the Indus River (Berg, 1949). This species ranges from Baluchistan to southern China and Java (Sufi, 1956).

The cyprinid genus *Garra* has representatives throughout most of the soudanian region of Africa and in suitable biotopes in Arabia, the Tigris and Euphrates system, Baluchistan, Pakistan and southern Asia from south China to Borneo (Menon, 1964). This genus has never been subjected to an adequate phylogenetic analysis, although in the only recent review Menon (*op. cit.*) attempted to group species into supra-specific complexes. He aligned *Garra rufa* with *Garra barriemiae* from Oman and the United Arab Emirates. Although the characters used for associating these two species have not been properly evaluated phylogenetically, the distribution of the two species is such that their association could well be valid. Today the saline barrier of the Persian Gulf and the arid areas of northern Arabia separate the two species. Formerly, however, these barriers did not exist. Until about 10 000 years ago, the Persian Gulf was filled with the freshwater discharge of the Tigris and Euphrates (Kassler, 1973) which, coupled with a wetter climate (Banister & Clarke, 1977) afforded the possibility of the complete occupation of the area encompassing the now discontinuous home ranges. However, it is still unknown to which other species *Garra rufa* and *Garra barriemiae* are related.

The loaches of the Tigris and Euphrates have been subject to even less study than have the cyprinids. The Spined Loach, *Cobitis taenia* is one of the most widely distributed Eurasian fishes, being found in suitable waters from Europe to Taiwan (Berg, 1964). *Sabanajewia aurata* is, by these standards, much more localized, occurring from the middle Danube and other Black Sea tributaries through Asia Minor to the Aral Sea drainage. Loaches of the genus *Noemacheilus* also occur throughout Eurasia and their absence from the Tigris and Euphrates would be of more significance than their presence.

The southern limit of distribution of the European catfish, the Wels *Silurus glanis* is the Tigris and Euphrates system. There is some doubt as to whether *Silurus triostegus* is specifically distinct from *Silurus glanis* and whether *Silurus chantrei* Sauvage occurs in the system. Sauvage (1882) described *S. chantrei* from Tiflis (Tbilsi) on the Kura River. Berg (1933, 1964) made the unsupported statements that *Silurus chantrei* is (a) a species of *Parasilurus* and (b) that '. . . its home, (is) allegedly the Kura River (actually Syria or the Tigris basin) . . .'. Haig (1952: 72) provided evidence that *Parasilurus* is not a valid genus but she retained *Silurus chantrei* as a distinct species; she also noted, but did not comment on, Berg's observations on the type locality. Unfortunately, there is no mention of *Silurus triostegus* in her revision. Berg (1949) records *Silurus triostegus* (as *Parasilurus triostegus*) from the Tigris and Euphrates. Hora & Misra (1943) put forward some arguments to suggest that *Silurus triostegus* is no more than a variant of *Silurus glanis*. They also wrote that *Silurus chantrei* 'is probably a synonym' of *Silurus asotus* L from China, Japan and eastern Russia. In view of the confusion surrounding this (or these) species no useful comments of a zoogeographical nature can be made.

The sisorid catfishes of the genus *Glyptothorax* and *Aclyptosternon coum* (see p. 97) from the Tigris and Euphrates are the westernmost representatives of their family. The other members of this family inhabit small rivers of southern and western Asia. Although species of the family Bagridae occur in both Africa and Asia, the single bagrid in the Tigris and Euphrates (*Mystus pelusius*) is a member of a genus limited to, but widespread in, southern Asia and peninsular India.

Thus, the catfishes occurring in the Tigris and Euphrates are representative of both European and southern Asian faunal elements.

It can be seen from the affinities of particular genera and species of primary freshwater fish discussed above (and on p. 104–106) that the Tigris and Euphrates have acquired a mixed fauna with a low endemicity.

This observation has been made by previous commentators on zoogeographical problems. DeBeaufort (1951), for example, regarded Mesopotamia as being in the holarctic, but also as belonging to a zone of transition between the holarctic and oriental regions. This transition zone has, as deBeaufort (*op. cit.*) pointed out, been regarded as a part of a larger (Tyrrhenian) zone incorporating holarctic-ethiopian transition regions as well as the holarctic-oriental transition region. Kuru (1971) evaluated the zoogeographical affinities of the 34 fish species he found in the Anatolian headwaters of the Tigris and Euphrates. He concluded that four elements were present: (1) western Palaearctic (European); (2) western Asian; (3) south-east Asian and Indian (with some African links – a Mesopotamian fauna); and (4) a Samartian (proto-Black Sea) component. Banarescu (1975) divided his Sino-Indian region (Oriental region, *auctt*) into four subregions. The most westerly (his west Asian region) includes the Tigris and Euphrates, the rivers of Syria, Lebanon, Israel, most of Iran, Afghanistan and a part of Pakistan. He regarded the west Asian region as possessed of a poor ichthyofauna, a result of the aridity of the region, and as having European ('Leuciscines', *Cobitis* and *Sabanejewia*) or Indo-Malayan (the 'Barbines') affinities. In particular, Banarescu decided, but did not attempt to substantiate his decision, that the genus *Barbus* had its origin in that region.

Kosswig (1956) suggested that the exchange of tropical and palaearctic faunas began in the Pliocene with the retreat of the Syrian-Iranian Sea (also known as the Fars Sea, the Sea having been named from its extensive deposits). With this sea in existence, the fauna of Anatolia was an entirely palaearctic one. Once the sea had retreated, Eremian (of desert origin) elements entered Anatolia from the south and east, and tropical elements via Jordan. Subsequently, the fauna of the Anatolian lakes was affected by glaciation.

As far as I can discover, no comprehensive reconstruction of the palaeogeography of this part of the world has been published. However, thanks to the kindness of Dr. G. F. Elliott of the British Museum (Natural History) the following reconstruction can be offered, based on unpublished information and his knowledge of the geology and geography of the region.

The Zagros orogeny started in the late Tertiary and continued as the Fars Sea dried up. The emerging Zagros mountains at first formed the eastern shore of the sea. The orogeny continued westwards, culminating in the uplift of the Anatolian plateau during the Pliocene. The uplift was partially responsible for the retreat and desiccation of the Fars Sea but as the plateau was uplifted lakes formed in the block faults and contained remnants of the Fars fauna (e.g. the mysid shrimp *Mesomysis*, see Kosswig, 1956). Some lakes dried out, leaving salt deposits, others were sumps for new rivers and gradually became less saline. The differing angles of plateau tilt, as the uplift continued, changed drainage direction and allowed many opportunities for river and lake capture. The Tigris and Euphrates would first have formed in the Pliocene when the water from the developing Zagros mountains drained away to the west and then flowed south into what is now the Persian Gulf. The establishment of the Anatolian highlands resulted in the northern extension of the Tigris and Euphrates as run-off increased the rate of back-cutting. During the Quaternary the lakes on the plateau were much larger than now and it is very likely that central Anatolia consisted of one very large lake (Lahn, 1948).

There is some evidence that the Mesopotamian region has long had a mixed fauna. Mecquenem (1924–1925) described early Pliocene mammal fossils from volcanic tuffs in north-west Iran and remarked that they had European, African and Asiatic affinities.

In the light of the geological history of the region, and in the absence of any phylogenetic studies on the fauna, the impossibility of arriving at any definitive conclusions on the zoogeographical affinities of the fishes of this region becomes understandable. The fauna has mixed origins, with substantial contributions from Europe as well as elements of a widespread, but generally Asiatic fauna and a smaller contribution from the rivers immediately to the east. The low endemicity may well reflect the unsettled nature of the pattern of water distribution, with few water bodies being isolated for a sufficient length of time to allow speciation to proceed.

Table. A compilation of all the names of species listed as occurring in the Tigris and Euphrates. In the left hand column (column 1) the genera and species are arranged alphabetically by families. The numbers in parentheses refer to the postscripts where the genera in question are discussed in detail. Column 2 contains the names by which the species in Column 1 are now known. A blank indicated that the name remains unchanged. Column 3 lists the authorities for the name changes.

Listed name	Current status	Authority
SALMONIDAE		
Salmo trutta L		
CYPRINIDAE		
[1] *Acanthobrama arrhada* Heckel	*Acanthobrama marmid*	Karaman 1972
Acanthobrama centisquama (Heckel)		
Acanthobrama marmid Heckel		
Acanthobrama orontis Berg	*Acanthobrama marmid*	Karaman 1972
Alburnoides bipunctatus fasciatus (Nordmann)		
[2] *Alburnus caerulus* Heckel		
Alburnus capito Heckel	*Chalcalburnus mossulensis*	Berg 1949
Alburnus mossulensis Heckel	*Chalcalburnus mossulensis*	Berg 1949
Alburnus orontis Sauvage		
Alburnus pallidus Heckel		
Alburnus schejtan Heckel	*Chalcalburnus mossulensis*	Berg 1949
Alburnus sellal Heckel	*Chalcalburnus sellal*	Berg 1949
[3] *Aspius vorax* Heckel		
[4] *Barbus barbulus* Heckel	*Barbus rajanorum*	Karaman 1971
Barbus belayewi Menon		
Barbus canis Valenciennes		
Barbus chantrei (Sauvage)	*Barbus canis*	Karaman 1971
Barbus esocinus (Heckel)		

Barbus euphrati (Sauvage)	Barbus esocinus	Karaman 1971
Barbus faoensis Günther	Barbus sharpeyi	Karaman 1971
Barbus grypus Heckel		
Barbus kersin Heckel	Barbus capito	Karaman 1971
Barbus kotschyi Heckel	Barbus grypus	Karaman 1971
Barbus lacerta Heckel	Barbus plebejus	Karaman 1971
Barbus longiceps Valenciennes		
Barbus lorteti Sauvage	Barbus longiceps	Karaman 1971
Barbus luteus (Heckel)		
Barbus mystaceus (Heckel)	Barbus rajanorum	Karaman 1971
Barbus orontis (Sauvage)	Barbus capito	Karaman 1971
Barbus pectoralis Heckel	Barbus capito	Karaman 1971
Barbus rajanorum Heckel		
Barbus scheich Heckel	Barbus rajanorum	Karaman 1971
Barbus scincus Heckel	Barbus plebejus	Karaman 1971
Barbus sharpeyi Günther		
Barbus subquincunciatus Günther		
Barbus xanthopterus (Heckel)		
Barilius mesopotamicus Berg	Leucaspius mesopotamicus	Howes (in press)
Barynotus albus (Heckel)	Barbus luteus	Karaman 1971
[5] Capoeta barroisi Lortet		
Chondrostoma nasus (L)		
Chondrostoma regium (Heckel)		
[6] Cyprinion kais Heckel	Cyprinion macrostomum	Berg 1949
Cyprinion macrostomum Heckel		
Cyprinion tenuiradius Heckel		
[7] Garra gymnothorax Berg	Garra rufa	Menon 1964
Garra lamta (Hamilton)		
Garra obtusa (Heckel)	Garra rufa	Menon 1964
Garra rufa (Heckel)		
Garra variabilis (Heckel)		
Hemigrammacapoeta nanus (Heckel)		
Leuciscus berak (Heckel)		
Leuciscus cephalus orientalis		
Leuciscus lepidus (Heckel)		
Leuciscus zeregi (Heckel)	Phoxinellus zeregi	Karaman 1972
Rutilus tricolor Lortet	Acanthobrama tricolor	Karaman 1972
Tylognathus elegans Günther	Hemigarra elegans	Karaman 1971
Typhlogarra widdowsoni Trewavas		
Varicorhinus damascinus (Valenciennes)	Capoeta capoeta	Karaman 1969
Varicorhinus trutta (Heckel)	Capoeta trutta	Karaman 1969
Varicorhinus umbla (Heckel)	Capoeta capoeta	Karaman 1969

COBITIDAE

Cobitis aurata (de Filippi)	Sabanajewia aurata	Banarescu et al.
Cobitis taenia L		1972
Noemacheilus angorae Steindachner		
Noemacheilus argyrogramma (Heckel)		
Noemacheilus frenatus (Heckel)		
Noemacheilus insignis (Heckel)		
Noemacheilus panthera (Heckel)		
Noemacheilus malapterurus (Valenciennes)		
Noemacheilus tigris (Heckel)		
Turcinonoemacheilus kosswigi Banarescu & Nalbant		

ARIIDAE

Arius cous see p. 97

Aclyptosternon coum
(SISORIDAE)   This paper

SISORIDAE

Glyptothorax cous see p. 97
Glyptothorax armeniacum Berg
Glyptothorax kurdistanicum Berg
Glyptothorax steindachneri (Pietschmann)

Aclyptosternon coum   This paper

SILURIDAE

Euglyptosternum coum see p. 97

Silurus glanis L.
Silurus triostegus Heckel

Aclyptosternon coum
(SISORIDAE)   This paper

BAGRIDAE

Mystus colvilli (Günther)
Mystus pelusius (Solander)

Mystus pelusius   Jayaram 1954

HETEROPNEUSTIDAE

Saccobranchus fossilis Valenciennes
Heteropneustes fossilis (Bloch)

Heteropneustes fossilis   Hora 1936

MASTACEMBELIDAE

Mastacembelus hallepensis Günther
Mastacembelus mastacembelus (Solander)

Mastacembelus mastacembelus Sufi 1956

---

[1] Species of the cyprinid genus *Acanthobrama* (*sensu* Goren *et alii*, 1973 not of Karaman, 1972) inhabit rivers from the eastern edge of the Mediterranean to the Tigris and Euphrates. The genus is not found east of the Tigris and Euphrates system. In the past there has been some confusion in the attribution of certain species to *Acanthobrama* or *Phoxinellus*. Karaman (1972) included the north African species *callensis* in *Acanthobrama*, but this species has now been shown by Howes (in press) to have been correctly placed in *Phoxinellus* by Pellegrin (1920). The presence of *Phoxinellus zeregi* (as *Leuciscus zeregi*) in the Tigris and Euphrates was noted by Al-Nasiri & Hoda (1976), but Karaman (1972) in his revision of *Phoxinellus* makes no mention of its presence in that river system. If, however, a species of *Phoxinellus* really does occur in the Tigris-Euphrates system then the genus has a most interesting distribution. As interpreted by Karaman (1972), *Phoxinellus* is a circum-Mediterranean genus. Subspecies of *zeregi* are found in Israel, Lebanon, Syria, Turkey and Azerbaijan. In Tunisia and Algeria are found *P. chaignoni* and *P. callensis*. Subspecies of *P. stimphalicus* and *P. adspersus* occur in Jugoslavia and Greece. *Phoxinellus pleurobipunctatus* is recorded from Greece and the island of Corfu. The significance of this circum-Mediterranean distribution remains obscure; perhaps a greater knowledge of the geomorphology of the region will increase our understanding.

[2] The degree of kinship between *Chalcalburnus* and *Alburnus* is unknown. Berg (1964) separated the two genera on the relative lengths of the ventral keel and the relative stoutness of the last unbranched ray in the dorsal fin, two characters which, at the 'generic' level, should be

treated with a great deal of suspicion in the absence of any corroborating evidence. As re-defined by Berg (1964) *Alburnus* has representatives throughout the European palaearctic region, whereas *Chalcalburnus* is confined to regions around the Black, Aral and Caspian seas.

(3) *Aspius vorax* is, by definition, most closely related to its only congener, the European *Aspius aspius*. A subspecies of the widespread European species (*Aspius aspius taeniatus*) is reported from rivers flowing into the southern part of the Caspian Sea (Berg, 1964). The two species can apparently be distinguished by the presence of more scales in the lateral line series of *Aspius vorax* (94–105 *fide* Beckman, 1962 *cf.* 65–74 for *Aspius aspius fide* Wheeler, 1969). However, according to Berg (1964) the Caspian subspecies has 67–90 scales in the lateral line series. Counts of the lateral line scales taken on specimens of *Aspius vorax* in the British Museum (Natural History) range from 93–101. It is therefore likely that the two species of *Aspius* may not be as morphologically distinct as had been thought. Indeed, the fact the Caspian subspecies is meristically, as well as geographically, closer to *Aspius vorax* than is the European subspecies suggests that a clinal phenomenon may be involved. Nonetheless, the presence of *Aspius* in the Tigris and Euphrates indicates a European influence on its fauna.

(4) The most speciose genus in the system is *Barbus*. *Barbus grypus* is a contender for the title of the largest fish in the Tigris and Euphrates; specimens nearly 2 m long and weighing 100 kg have been reliably reported (Beckman, 1962 and Elliott, 1977). *Barbus* presents many difficulties to systematists. The species are morphologically very variable, and it is this variability, whether it be predominantly genotypically or phenotypically controlled, that has led to the description of many nominal species. Only when extremely large series of specimens have been studied can we have any degree of confidence that the taxonomist's species bear any relation to those in nature. The variability, both intra- and interspecific, may be so marked as to lead workers into erecting new genera (see p. 98). The majority of the *Barbus* species in the Table can be divided into two stocks, each of which may be monophyletic. These are the 'European' stock (fishes with a cylindrical body, small scales and a serrated dorsal spine) and the 'Afro-Indian' stock (fishes with a compressed body, large scales and a smooth dorsal spine). Of the 12 recognized species in this region, seven (*B. belayewi*, *B. esocinus*, *B. longiceps*, *B. plebejus*, *B. rajanorum*, *B. subquincunciatus* and *B. xanthopterus*) belong to the 'European' stock. The species *B. capito*, *B. grypus* and *B. sharpeyi* have the characters of the 'Afro-Indian' stock. Parenthetically, it may be mentioned that some of the Indian species of the 'Afro-Indian' stock, the Mahseers, are characterized by their large size, a trend noticeably manifest in *B. grypus*. The two remaining species, *Barbus luteus* and *B. canis* are less satisfactorily placed in a complex. Although superficially like the 'Afro-Indian' stock species, they differ in having six, not five, branched rays in the anal fin. The derived condition of the extra ray in the anal fin, along with the characters typical of the 'Afro-Indian' stock also occurs in the Arabian species *Barbus exulatus* and *B. apoensis* and it has been argued that the four species are closely related (Banister & Clarke, 1977).

Overall, the zoogeographical affinities of the *Barbus* species of the Tigris and Euphrates are mixed, apparently containing endemic species as well as Afro-Indian and European components.

(5) Members of the cyprinid taxon *Capoeta* have a wide ventral mouth and the lower jaw covered by a sharp-edged 'horny' sheath. The pharyngeal teeth have characteristic horseshoe-shaped crowns. Following the revision of Karaman (1969) the genus includes species formerly placed in the African genus *Varicorhinus*. *Capoeta* is endemic to the region from northern Asiatic Turkey to northern Afghanistan (both the internal basins and the Aral drainage, Karaman, 1969). *Capoeta capoeta* is the most widespread species, its range encompassing those of all the other species. The relationships of *Capoeta* to any other cyprinids are unknown.

(6) The cyprinid genus *Cyprinion* is endemic to the freshwaters between Syria and India, that indefinable region often loosely called the Middle East. Berg (1949) revised the genus and concluded that he could identify some supra-specific complexes. His most speciose group, the '*C. watsoni*' complex is characterized by the possession of less than eleven branched rays in the dorsal fin, and occupies the east and north-eastern parts of the generic range. Neither of the species alleged to be present in the Tigris and Euphrates belongs to this group (they have more

than 12 branched rays in the dorsal fin) so their affinities cannot lie with the eastern species. The relationships of the genus as a whole, and of *Cyprinion macrostomum* and *C. tenuiradius* in particular, are unknown.

It is possible, however, that the present-day distribution of the genus is less extensive than it was formerly. Hora (1956) described fish paintings on pots from the third millennium B.C. from Lal in Baluchistan, and discussed the zoogeographical implications of the genera depicted. The quality of the paintings is such that many of the genera depicted can be identified beyond doubt; the identification of other genera represented is, however, less certain. One of these is the fish identified by Hora as a *Cyprinion*. Here, let me say that I think Hora's identification is the most likely; it is impossible to be dogmatic on this issue, but there are no other extant genera with which the fishes in these paintings could have been confused, especially if the standard of accuracy is constant throughout all the pictures. Hora argues that the significance of these discoveries is that *Cyprinion* has become extinct in that region since the third millennium B.C. and formerly the rivers were more extensive, thus indicating a wetter climate and an enlarged ichthyofauna.

(7) Menon (1964) regarded *Garra variabilis* as related to *Garra rossica* from Afghanistan and he thought that these two species represents 'the most primitive known group of species within the genus'. However, the validity of relationships based on shared primitive characters has recently been questioned, so for the moment the postulated relationship of *Garra rossica* and *Garra variabilis* must be left in abeyance. The record of *Garra lamta* in the Tigris and Euphrates is regarded with some suspicion (see p. 97). If Menon's (*op. cit*) conclusions are correct, then the *Garra* species of the Tigris and Euphrates are each, and separately, more closely related to species from the east and west than they are to one another.

Whilst considering the genus *Garra*, mention should be made of the blind, hypogean species *Typhlogarra widdowsoni*. Implicit in the choice of generic name for this species is a close relationship with the epigean *Garra*. *Typhlogarra* is found in underground streams near Haditha, the waters of which ultimately drain into the Tigris and Euphrates. In the wetter climate pre 10 000 B.P., discussed above, it is likely that surface streams existed where there are now only subterranean ones. It might have been thought with only two epigean *Garra* spp. in the region the sister group (or groups) of the eyeless species would be easy to determine. Recent researches (which are far from complete) have suggested that this is not so. Firstly, and rather surprisingly for a cavernicolous fish, *Typhlogarra* has massive, enlarged pharyngeal bones with only three large, kidney-shaped, molariform teeth forming the inner row. The second row is usually absent, although in one specimen there are two minute teeth present. There is no trace of the third row. All the *Garra* species so far investigated have three rows of blade-like teeth arranged in a 5.3.2 pattern (fig. 28 in Banister & Clarke, 1977). In both *Garra rufa* and *Garra variabilis* the pharyngeal bones are relatively slender and fragile. Both *Garra fura* and *Typhlogarra* have four barbels and a mental disc. *Garra variabilis* has only two barbels. To confuse the issue further, two apparently distinct species inhabit the same underground system, the second species, however, has a slender pharyngeal bone and lacks the mental disc. On the basis of the characters so far examined, the new species is much more closely related to the epigean *Garra* spp than is *Typhlogarra*. The presence of *Typhlogarra* does nothing to aid any zoogeographical considerations on *Garra*.

References to Chapter 8

Al-Hamed, M. I. 1966. Limnological studies on the inland waters of Iraq. Bull. Iraq. nat. Hist. Mus. *3* (5): 1–22.

Al-Nasiri, S. K. & Hoda, S. M. S. 1976. A guide to the freshwater fishes of Iraq. Basrah nat. Hist. Mus. Publ. (1): 1–125.

Banarescu, P. 1973. Some reconsiderations on the zoogeography of the Euro-Mediterranean fresh-water fish fauna. Revue Roum. Biol. *18* (4): 257–264.

Banarescu, P. 1975. Principles and problems of zoogeography. pp. 214, U.S. Department of Commerce, National Technical Information Service, Springfield Va 22151. TT 71-56006.

Banarescu, P., Nalbant, T. T. & Chelmu, S. 1972. Revision and geographical variation of *Sabanajewia aurata* in Romania and the origin of *S. bulgarica* and *S. romanica* (Pisces, Cobitidae). Annot. zool. bot. Bratislava (75): 1–49.

Banister, K. E. & Clarke, M. A. 1977. The freshwater fishes of the Arabian peninsula. In: The scientific results of the Oman flora and fauna survey 1975. Jl. Oman Stud. 1977: 111–151.

Banister, K. E. & Clarke, M. A. (In Press). A revision of the large *Barbus* of Lake Malawi with a reconstruction of the history of the southern African rift valley lakes. J. nat. Hist.

De Beaufort, L. F. 1951. Zoogeography of the land and inland waters. Sidgwick & Jackson, London. 208 pp.

Beckman, W. 1962. The freshwater fishes of Syria and their general biology and management. 4° Rome, U.N. F.A.O. fisheries division technical paper No. 8: 297 pp.

Berg, L. S. 1933. Les poissons des eaux douces de l'URSS et des pays limititrophes. 3rd edn, revue et augmentée. *1* Leningrad, 543 pp.

Berg, L. S. 1949. Freshwater fish of Iran and of neighbouring countries. (In Russian). Trudy zool. Inst. Leningr. *8* (4): 783–858.

Berg, L. S. 1964. Freshwater fishes of the U.S.S.R. and neighbouring countries. *2* (English translation of the 4th edition of 1949), 496 pp.

Bleeker, P. 1863. Systema silurorum revisum. Ned. Tyd. Dierk *1*: 77–122.

Elliott, G. F. 1977. Untitled letter. Proc. Geol. Ass. *87*: 430.

Fang, P. W. 1943. Sur certaines types peu connu de cyprinides des collections du Museum du Paris. Bull. Mus. natn. Hist. nat. Paris (2) *15*: 399–405.

Goren, M., Fishelson, L. & Trewavas, E. 1973. The cyprinid fishes of the genus *Acanthobrama* Heckel and related genera. Bull. Br. Mus. nat. Hist. (Zool.) *24* (6): 293–315.

Gray, J. E. 1834.* The illustrations of Indian zoology, chiefly selected from the collections of General Hardwicke. London. (2).

Gunther, A. 1868. Catalogue of the fishes in the British Museum 7. Trustees, British Museum, London xx + 512 pp.

Haig, J. 1952. Studies on the classification of the catfishes of the oriental and palaearctic family Siluridae. Rec. Indian Mus. *48* (3–4): 59–116.

Heckel, J. J. 1843. Ichthyologie. In: Russeger, J. Reisen in Europa, Asien und Afrika *1*: 991–1099. Stuttgart.

Hora, S. L. 1923. Notes on fishes in the Indian museum, 5. On the composite genus *Glyptosternon* McClelland. Rec. Indian Mus. *25* (1): 1–44.

Hora, S. L. 1936. Siluroid fishes of India, Burma and Ceylon. Rec. Indian Mus. 38(2): 199–209.

Hora, S. L. 1956. Fish paintings of the third millennium B.C. from Lal (Baluchistan) and their zoogeographical significance. Mem. Indian Mus. *14* (2): 73–86.

Hora, S. L. & Misra, K. S. 1943. On a small collection of fish from Iraq. Jl R. Asiat. Soc. Beng. *9* (1): 1–15.

Howes, G. J. 1978. The anatomy and relationships of the cyprinid fish *Luciobrama macrocephalus* (Lacepede) Bull. Br. Mus. nat. Hist. (Zool.) *34* (1): 1–64.

Howes, G. J. (In press). The anatomy, phylogeny and classification of the bariliine cyprinid fishes. Bull. Br. Mus. nat. Hist. (Zool.).

* 1833 is the date customarily attributed to this work, but it was issued over a period of 5 years in 20 parts. The dating of the parts has been authenticated by Sawyer (1958) and the name *Tor* appeared in 1834 in parts 17 and 18.

Jayaram, K. C. 1954. Fishes of the genus *Mystus* Scopoli. Rec. Indian Mus. *51* (4): 527–558.

Karaman, M. S. 1969. Revision der kleinasiastischen und vorderasiatischen Arten des Genus *Capoeta* (*Varicorhinus*, partim). Mitt. hamb zool. Mus. Inst. *66*: 17–54.

Karaman, M. S. 1971. Revision der Barben Europas, Vorderasiens und Nordafrikas. Mitt. hamb. zool. Mus. Inst. *67*: 175–254.

Karaman, M. S. 1972. Revision einiger kleinwüchsiger Cyprinidengattungen *Phoxinellus*, *Leucaspius*, *Acanthobrama* usw, aus Sudeuropas, Kleinasien, Vorder-Asien und Nordafrika. Mitt. hamb. zool. Mus. Inst. *69*: 115–155.

Kassler, P. 1973. The structural and geomorphic evolution of the Persian Gulf. In: The Persian Gulf. Ed. Purser, B. H. Springer-Verlag, Berlin: 11–33.

Khalaf, K. T. 1962. The marine and freshwater fishes of Iraq. Ar-Rabitta Press, Baghdad. 164 pp.

Kosswig, C. 1956. Zoogeography of the Near East. Syst. Zool. *4* (2): 49–73.

Kuru, M. 1971. The fresh-water fish fauna of eastern Anatolia. Istanb. Univ. Fen. Fak. Mecm. Ser. B *36* (3–4): 137–146.

Ladiges, W. 1964. 3 Teil, restliche Gruppen. Mitt. hamb. zool. Mus. Inst. *61*: 203–220.

Lahn, E. 1948. Contribution à l'étude géologique et géomorphologique des lacs de la Turquie. Maden Tetik Arama Enstit. Yayinl. (B) *12*: 89–178.

LeStrange, G. 1905. The lands of the eastern Caliphate: 26–27. Cambridge Geographical Series. xvii + 536 pp.

Lees, G. M. & Falcon, N. L. 1952. The geographical history of the Mesopotamian plains. Geogr. J. *118*: 24–39.

Mahdi, N. 1962. Fishes of Iraq. Ministry of Education, Iraq. 82 pp.

Mahdi, N. & Georg, P. V. 1969. A systemic list of the vertebrates of Iraq. Iraq nat. Hist. Mus. Publs (26): 1–104.

Mecquenem, R. de. 1925. Contribution à l'étude des fossiles de Maragha. Annls Paléont. *13*: 135–160; *14*: 1–36.

Menon, A. G. K. 1964. Monograph of the cyprinid fishes of the genus *Garra*, Hamilton. Mem. Indian Mus. *14*: 173–260.

Pellegrin, J. 1920. Sur deux cyprinides nouveaux d'Algerie et de Tunisie appartenent au genre *Phoxinellus*. Bull. Mus. natn. Hist. nat. Paris (1) *26* (5): 372–375.

Sauvage, H. E. 1882. Catalogue des poissons recueillis par M. E. Chantre pendant son voyage in Syrie, Haut-Mesopotamie, Kurdistan et Caucase. Bull. Soc. philomath. Paris (7) *6*: 163–168.

Sawyer, F. C. 1953. The dates of issue of J. E. Gray's 'Illustrations of Indian zoology' (London, 1830–1835). J. Soc. Biblphy nat. Hist. *3* (1): 48–55.

Sufi, S. M. K. 1956. Revision of the oriental fishes of the family Mastacembelidae. Bull. Raffles Mus. (27): 93–146.

Wheeler, A. C. 1969. The fishes of the British Isles and north west Europe. Macmillan, London, 613 pp.

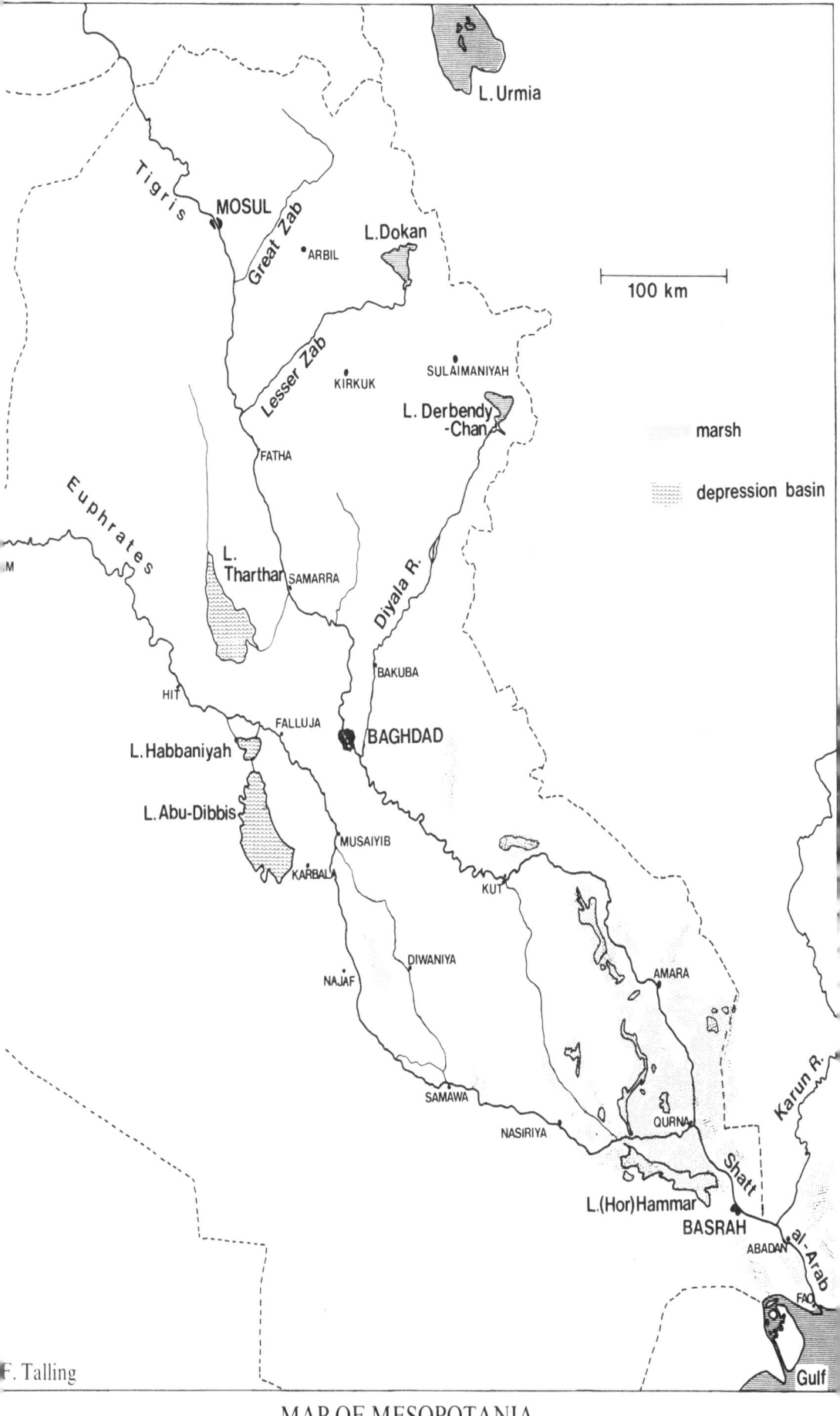

MAP OF MESOPOTANIA

Epilogue

I have tried in this book to assemble some outstanding features of the scene of Mesopotamia, a name coined by the ancient Greeks and still ecologically sound. I am aware of the fragmentary nature of this presentation; an enormous literature exists on the past of this land but not comprehensive for the natural history of the present. I enclose here a summary of the main points of the interdisciplinary approach.

Summary

(1) The present political areas of the Near East are part of a wider scene of similar ecological features. Geological, morphological and climatic factors have contributed to desert, mountains of a wide semicircle around arid and often saline lands.

(2) Mesopotamia is the result of the unique configuration of a river-system which descends southwards to the sea. Orogenic earth movements in the Tertiary created mountains in the north and east, which gathered water of snow-melt and rain and forced the incipient streams to flow into the synclinal low plain of Mesopotamia.

(3) The resulting two great rivers, Euphrates and Tigris with their tributaries, thus began their work of the transport of water and erosion products, covering the low plain with alluvial soils and a continuous supply of water. In the west the Arabian desert shield, an old formation hemmed in the alluvial valley.

(4) Of the present Mesopotamian Iraq 67 per cent is covered by desert-steppe, 18 per cent by hills and mountains and the remaining 25 per cent by alluvial soils. With such distribution the climate decides the distribution of plant cover as basis for life. Winters are cold and humid, summers hot and practically rainless. Short lived annuals may spring up in the desert with few draught-resisting shrubs; with increasing rainfall towards the northern and eastern hills and mountains, a much richer vegetation exists. The natural plant-cover is now severely reduced by the ravages of man and in the last ten thousand years by increasing aridity.

(5) Archaeologists have not only described the rise of civilizations in Mesopotamia but have revealed the much more favourable palaeo-ecology of the Near East in the palaeolithic and up to the neolithic conditions leading to the beginnings of agriculture and irrigation practices, making life possible in the plains along the rivers.

(6) The rivers dominate the landscape of Mesopotamian Iraq. Thus their regime, slopes, discharge and currents play a decisive role. Spring floods from the mountains increase the volume of water drastically, and often disastrously with a strong fall in summer levels; slopes influence the currents and their capacity of carrying and depositing sediments. They

also diminish the force of the rivers and both the Euphrates and the Tigris show a slow down in the alluvial plain with meandering of their courses. After descending from the mountains the rivers of Mesopotamia wind their way through a mostly flat land and form many inundation basins, which are extensive but shallow, and often the rivers split up into small channels with the development of marshes. These are extensive in the southern reaches and form a characteristic habitat for life, including man.

(7) The outstanding feature of the Mesopotamian river-system is its instability and far reaching and often radical changes have been recorded since antiquity. A special case is the much disputed advance of the Delta formed by the interaction of the two main rivers into the Shatt al-Arab, and of the Karun river from Iran.

(8) All these factors influence the vegetation and grossly diminished fauna of the land and the life in waters. Dr. J. F. Talling has surveyed the water characteristics and the resulting phytoplankton of the waters of Iraq. The influence of saline sub-soils, deposited by a long sequence of sea transgression from the Mesozoic until the Tertiary, influences the chemical set up of all waters not only in Iraq, but also in neighbouring areas of Palestine, Jordan, Syria and Iran. Irrigation practised in the Near East for at least 6000 years, brings the salt solutes into the waters.

(9) What is known of the aquatic fauna is fragmentary and has been collated. Dr. K. E. Banister has demonstrated this on the example of fishes, very imperfectly known. Large lacunae exist in our knowledge in the aquatic fauna of Iraq and have to be filled by local scientists, now organized in a Biological Research Centre at Baghdad.

General and personal reflections

There is now a renewed interest in long rivers after some years of less attention. Early valuable investigations in Europe and America, 50 years ago, are now renewed and monographs are beginning to appear. By their length the great rivers present considerable technical difficulties, which can only be solved by team work and interdisciplinary approach. Almost all rivers are closely connected with man; the Nile, Euphrates and Tigris, the Indus and the Yellow river have been the focus of past civilizations, to name the best known. At present many rivers have been harnessed by man with extensive use of the science of hydrology and hydraulics. There is a great difference between lakes and rivers, the first 'work' mainly internally by biological production and closed sedimentation, rivers 'work' by the force of their flow. Rivers form and change landscapes, collect influences from afar and drain the surface of their lands. They have attracted early man as communication routes and thus influenced interchanges of peoples, cultural and economic values. A study of rivers cannot be confined to particular items of hydrochemistry or particular morphological or biological features, though such work is essential. But ultimately all the various features are and have to be joined, if the nature of these great arteries is to be understood.

Personal remarks

I must confess that my interest of collating this book, arose from the close connection of history and river action. I saw the great Jordanian desert stretching towards the Mesopotamian valley, from the air I contemplated the arid lands of Arabia to the western Pakistan. This book has been written therefore mainly from a 'bird's eye' view and as the space photographs show even from a space view. This brings out the generalities and omits many details.

I hope that I have avoided rash opinions or what a German scientist years ago called 'das Problem der Problem Verderbung' indicating a premature and imperfect rendering of a great natural phenomenon.

Annexe: Mesopotamian past as seen by eyewitnesses

The chapter on palaeo-ecology could be enlarged by a diligent search of the vast array of existing sources and such enquiry would add to the reconstruction of the last man-dominated phase of this ancient land. Such attempt is beyond my present possibilities. Only some few examples of the accounts of eyewitnesses can be quoted; a great amount of observations by the ancients recorded on stone and clay tablets is incorporated in books on Sumer, Assyria and Babylonia.

Eyewitnesses are mostly travellers and their accounts are a reflection of personal and time bound curiosity; interest in the environment has changed with the ages. Personal curiosity and perceptiveness is even a stronger filter of accounts as seen by comparing the quotation from *Herodotus* and the hurried remarks by Marco Polo. The history of Mesopotamia was 3000 years old when Herodotus visited the country in the wake of Greek penetration and recorded his astute impressions in the 4th century B.C. *(Histories; book I)*, quoted after the new translation by Aubrey de Selincourt 1965.

... Cyrus, having subdued the rest of the continent, ... turned his attention to Assyria, a country remarkable for the number of great cities it contained, and especially for the most powerful and renowned of them all – Babylon, to which the seat of government was transferred after the fall of Nineveh. Babylon lies in a wide plain in the form of a square with ... a circuit of some fifty six miles ... surrounded by a deep moat ... [the excavated] soil was formed into bricks, which were baked in ovens ... Using bitumen for mortar ... they laid rushmats between every 30 courses of brick ... Eight days journey from Babylon there is a city called Is on a small river of the same name, a tributary of the Euphrates, and in the river lumps of bitumen are found in great quantity. The Euphrates, a broad, swift river, which rises in Armenia and flows into the Persian Gulf, runs through the middle of the city ... [which has] towers which can be climbed by a spiral way running on the outside ... On the summit is a great temple ... contemporary inhabitants, Chaldaeans, speak about sacrifices and gold statues ... but I never saw it myself. ... [Two queens] were responsible ... for remarkable embankments in the plain ..., built to control the river, which until then used to flood the whole countryside ... [Another queen] changed the course of the river from straight to a winding one for security ... [and] dug a basin for a lake some 47 miles in circumference ... the purpose was to cause the frequent bends to reduce the currents. ...

The rainfall of Assyria is slight and provides enough moisture only to burst the seed and start the root growing, but to swell the grain and bring it to maturity artificial irrigation is used, not as in Egypt by natural flooding of the river but by labourers working hand-pumps. Like Egypt the whole country is intersected by dykes ... the largest of them ... runs in a south-easterly direction until it joins another river, the Tigris, on which Nineveh was built. As a grain-bearing country Assyria is the richest of the world ... Yields of wheat and barley are 200-fold and in exceptional years 300-fold, the sizes of millet and sesame are great, who was not been in Babylon will not believe it . . .'.

Herodotus was surprised by the river traffic, especially by the coracles which came to Babylon 'from Armenia' up to '14 tons in size, built of skins stretched over a wooden frame, floating down with the strong current and with 2 men and a donkey on board. At Babylon they are dismantled, the cargo and the wood sold, the skins carried back overland on the donkey. *Four rivers run across Armenia, the Tigris, two called Zabatu, the fourth is the Gyndes'*, the present Karun. – There are further observations on fruit trees, the datepalms and the life of the people, their dresses and rituals, fish eating clans and other matters, which struck Herodotus as worth noting. In general this is a remarkable record of shrewd observations including river limnology.

Three centuries later *Strabo*, who lived from 63 B.C. to 23 A.D., wrote his *'Geography'*, translated by H. L. Jones (Loeb Classical Library 1917 onwards). I have not read his account of Mesopotamia and I can quote only his remarks on the delta of the Mesopotamian rivers, mentioned in the Handbook of Iraq and the Persian Gulf (p. 54/55). Strabo writes of the barrier created by the advance of the Kharkey-Karun rivers; this is now believed to have occurred between 500 and 300 B.C. The barrier caused *'the overflow of the water, falling into the plains by the sea, making lakes, marshes and fens'*. He disputes earlier assumptions of Erathostene that the marshes drain by percolation into Syria because of mountains and writes *'These marshes are near the Persian Sea and the isthmus which separates them is not wide or rocky'*. Obviously Strabo was there and is a true eyewitness.

By the end of the 13th century *Marco Polo* travelled through Mesopotamia on his mercantile and christianizing mission to the great Khan of Mongolia and China. He came from the great trading centres of Genoa and Venice, which seemed to keep the monopoly of the then most precious imports, spices and silk. He apparently passed through Mosul and notes in his christian fervour the different sects of Nestorians, Jacobites and Armenians; he does not like the 'Infidels' in his case the Moslems. But he reverts quickly to 'cloths of gold and silk' and 'the large quantities of spices and drugs' traded by Greek merchants. *'In the mountainous parts there is a race of people named Kurds, some Christians . . . others Mahometans . . . Baudas (Baghdad) is a large city . . . a great river flows through the middle, by which the merchants transport their goods to and from the Indian Sea, the distance being computed at seventeen days of navigation, in consequence of the windings of the course . . . Before reaching the anchorage they pass a city named Balsara [Basra], in the vicinity of which are groves of palm trees producing the best dates in the world . . .'.* (Travels of Marco Polo, edited by Milton Rugoff, New American Library 1961, chapters 5 and 6).

Religious interest in the Bible stories of 'The Flood', the Jewish exile and 'Nineveh the accursed' drove travellers through the arduous desert routes, mostly via Aleppo. Some of their itineraries are briefly mentioned by Seton Lloyd (1955). Frenchmen, Italians, Spanish Jews, and Britons succeeded each other especially from the 16th century onwards. Only few noticed the shapeless mounds, hiding the remains of past civilizations. The Ottoman rule lay heavily and indolently on the Near East, the population was oppressed and lethargic. Many irrigation installations had fallen into ruin, taxation was excessive and banditry was rife.

By the late 18th but more intensively in the 19th century a new type of explorer came on the scene of Mesopotamia in search of the past, the archaeologists. Two nations were in sharp competition at first, the French and the British, later joined by Americans, Germans and others. Most of them left not only the scholarly descriptions of their excavations but also travelogues. Their results, experiences, difficulties are brought together by Seton Lloyd (1955). It would be fascinating to collate their descriptions of the land of Mesopotamia and its state in the 19th century. Here we can only mention some very general impressions, gained by them while traversing the country on camel or horse. Heat, dust storms, the overwhelming arid desert and the contrast to the riverine green cultivations, nomad tribes in revolt, the instability of the Turkish administration and the resulting deterioration of the land are the main themes. But the ardour of discovering and digging up the traces of the rich civilizations of Assyria, Babylonia and later of Sumer fired archaeologists in increasing numbers until the present. Guest (1966) describing the present state of the vegetation of Iraq has remarked on the 'degradation' visible in many parts (pp. 80–82) and especially the forest zone in the hills and mountains. He quotes (pp. 85–89) observations of travellers in the 19th century noting the deforestation and adds that, in spite of present regeneration work, many parts still show the neglect of centuries.

The influence of man on the quality of the land is, of course, visible all over the world; e.g. in Britain vegetation and the fauna have changed even over the last two hundred years, similarly in the United States. In Mesopotamia man has been active at least 6000 years and even before.

And then is the desert pressing relentlessly on man's creative efforts. The last eyewitness quoted here is the archaeologist *L. Woolley*, who in his book 'Ur of the Chaldees' (1936) wrote in a melancholy mood: . . . *'Only to those who have seen the Mesopotamian desert will the evocation of the ancient world seem well-neigh incredible, so complete is the contrast between past and present. The transformation of a big city into a tangle of shapeless mounds shrouded in driftsand and littered with broken pottery is not easy to understand, but it is yet more difficult to realise, that the blank waste ever blossomed, bore fruit for the sustainance of a busy world. Why, if Ur was an empire's capital, if Sumer was once a vast granary, has the population dwindled to nothing, the very soil lost its virtue?'*

The immediate answer is the withdrawal of the Euphrates to a new course east, much farther than the then water transport abilities. The more general answer is the universal passing of civilizations. The testimony of eyewitnesses of the past enables us to assess the present efforts of reconstruction.

References to Chapters 1–5 and 7

References to Chapters 6, 6a and 8 are included there

Allouse, B. E. 1959. A handlist of the birds of Iraq. Iraq Nat. Hist. Museum, publ. no. 2 Baghdad.

Anopheles and Malaria in the Near East. 1950. London School of Hygiene and Tropical Medicine, Memoir 7. See also under Macan.

Baghdad Nat. Hist. Museum. Number of bulletins amongst them bull. 6: Bibliography of Iraq and neighbouring countries.

Bonavia, S. 1894. The Flora of Assyrian Monuments. London.

Boxhall, G. A. 1976. A new genus and two new species of Copepods parasitic on freshwater fishes. Bull. Brit. Mus. Nat. Hist., Zoology 30: 209–213.

Braidwood, R. J. and B. Howe. 1960. Prehistoric Investigations in Iraqi Kurdistan. Orient. Instit. of the Univ. of Chicago. Studies in Ancient Oriental Civilisations No. 31. Univ. of Chicago Press.

Brinkmann, W. L. F. 1967. Grundlagen der Bewässerungswirtschaft am mittleren Euphrat. Zeitschr. f. Bewässerungswirtschaft 1: 65–77 (2 more papers ibidem 1968).

Buringh, P. 1957. Living conditions in the lower Mesopotamian plains in ancient times. Sumer 13: 30–46.

Buringh, P. 1960. Soils and soil conditions, Baghdad, Govt. publication.

Buringh, P. and L. Kadry. 1956. Soil survey and classification. Proc. 6th Int. Congr. of Soil Science, Paris 5, 14: 83–86.

Butzer, K. W. 1972. Environment and Archaeology. An Ecological Approach to Pre-history. Methuen, London (2nd edition).

Butzer, K. U. 1965. Physical Conditions in Eastern Europe, Western Asia and Egypt before the period of agricult. and urban settlement. In: Cambridge Ancient History. Rev. ed. Vols. I and II.

Buxton, P. A. 1923. Animal Life in Deserts, a study of the fauna in relation to the environment. London, E. Arnold.

Chapman, V. J. 1960. Salt Marshes and Salt Deserts of the World. Plant Sc. Monogr. ed. by N. Polunin. London, L. Hill Books.

Cheesman, R. E. 1920. Report on the mammals of Mesopotamia. J. Bombay Nat. Hist. Soc. 27.

Chopard, L. 1920. Report on the Orthoptera of Mesopotamia and Persia. J. Bombay Nat. Hist. Ser. 27: 759–771.

Cressey, G. B. 1958. The Shatt al-Arab basin. Middle East J. 12: 448–460.

Darshshuri, B. P. 1978. Threatened cats of Asia, the Asiatic cheeta. Wildlife Magazine, London.

Fedden, R. 1965. Syria and Lebanon, London, J. Murray.

Gabaly, el M. M. 1976. Problems and Effects of Irrigation in the Near East region. In: Arid Zone Irrigation Symposium. Alexandria 1976, E. Worthington ed.

Geographical Handbook. Naval Intelligence Division. 1944. Geograph. Handbook Series. Iraq and the Persian Gulf. London.

Guest, E. 1966. Flora of Iraq vol I Introduction. Min. of Agric. publ. Baghdad. University Press Glasgow, E. MacLehose & Co.

Gurney, R. 1920. Freshwater Crustacea coll. by Dr. P. A. Buxton in Mesopotamia and Persia. J. Bombay Nat. Hist. Soc. 27: 835–843.

Hamad, Al-M. 1976. Limnological investigations on Dokan reservoir. Bull. Nat. Hist. Centre, Baghdad, 7: 91–109.

Handel-Mazzetti, H. 1914. Die Vegetationsvehältnisse von Mesopotamien. Wiss. Ergebnisse der Expedition nach Mesopotamien 1910 Annal. Naturhistorisches Hofmuseum, Wien 28: 1–65.

Hawkes, Jacquetta, ed. 1963. The World of the Past. Introduction and introductory notes by J. Hawkes. New York, A. Knopf.

Hawkes, Jacquetta. 1973. The First Civilisations. Life in Mesopotamia, the Indus valley and Egypt. London, Hutchinson.

Helbaek, H. 1960. The ecological effects of irrigation in ancient Mesopotamia. Iraq vol. 22: 186–196.

Helbaek, H. 1960. The palaeobotany of the Near East and Europe. In: Braidwood and Howe: 99–117.

Helbaek, H. 1966. Commentary on the phylogeny of Triticum and Hordeum. Economic Botany 90: 350–360.

Helbaek, H. 1969. Plant collecting, dry farming and irrigation agriculture in pre-historic Deh Luran. In: Mem. Mus. Anthrop. Univ. Michigan 1, app. 1: 383–426.

Helbaek, H. 1972. Traces of plants in the early ceramic site of Umm Dabaghiyah. Iraq, Brit. Archaeol. School, 34: 17–19.

Herodotus. The Histories. Newly translated and with an introduction by Aubrey de Selincourt 1965. Penguin Classics, England.

Ionides, M. G. 1937. The Regime of the Rivers Euphrates and Tigris. London, Spon Ltd.

Ionides, M. G. 1955. Reply to Lees and Falcon. Geogr. J. 121: 394–395.

Ionides, M. G., S. Smith et al. 1954. The geographical history of the Mesopotamian Plain, Geogr. J. 120: 394–397.

Jacobsen, Thorkild. 1960. The waters of Ur. Iraq vol. 22: 174–185.

Jacobsen, The & Adams 1958. Salt and silt in ancient Mesopotamia. Science 128: 1251.

Khalaf, A. N. 1959. (Notes on Reptilia). Baghdad Nat. Hist. Museum, Bulletin.

Khalaf, A. N. and N. N. Smirnov. 1976. The littoral Cladocera of Iraq. Hydrobiologia, 51: 91–94.

Larsen, C. H. 1975. The Mesopotamian Delta region, a reconsideration of Lees and Falcon. J. American Oriental Society, 95: 43–57.

Lees, G. M. and N. L. Falcon. 1952. The geographical history of the Mesopotamian Plain. Geogr. J., 118: 24–39.

Lloyd, Seton. 1955. Foundations in the Dust. Pelican Books, Gt. Britain.

Lloyd, Seton. 1960. Ur, Al-Ubaid, Uqair, and Eridu. Iraq (commemor. vol. for Woolley), 22: 23–31.

Löffler, H. 1956. Limnologische Untersuchungen an Iranischen Binnengewässern. Hydrobiologia vol. 8.

Löffler, H. 1959. Beiträge zur Kenntniss der Iranischen Binnengewässer. 1. Der Nirizsee und sein Einzugsgebiet. Int. Rev. Hydrobiologie, 44: 227–276.

Löffler, H. 1961. Beiträge zur Kenntniss der Iranischen Binnengewässer. 2. Regional-limnologische Studie mit besonderer Berücksichtigung der Crustaceenfauna. Hydrobiologia, 17: 309–406.

Macan, T. T. 1950. The anopheline mosquitoes of Iraq. In: Anopheles and Malaria etc.

Mallowan, M. E. L. 1960. Ur in retrospect. In memory of Sir C. Leonard Woolley. Iraq 22. Published by Brit. School of Archaeology in Iraq.

Meissner, B. 1891. Babylonische Pflanzen-namen. Zeitschr. für Assyriologie 6.

Mohammad, M. M. 1965. A faunal study of the Cladocera of Iraq. Bull. Biol. Res. Centre, Baghdad, 1: 1–11.

Morice, F. D. 1921. Annotated lists of aculeate Hymenoptera . . . recently coll. in Mesopotamia and N.W. Persia. J. Bombay Nat. Hist. S. 27: 816–828.

Morton Boyd, J. 1967. Intern. Jordan Expedition. Report, roneographed. Conserv. Section of Intern. Biol. Programme, Nature Conservancy, London.

Oates, Joan. 1960. Ur and Eridu, the prehistory. Iraq, vol. 22: 32–50.

Parrot, A. 1961. Nineveh and Babylon. Transl. by S. Gilbert and J. Emmons. In series: Arts of Mankind ed. by A. Malraux and G. Salles. Engl. edition by Thames & Hudson.

Peile, H. D. 1921. Butterflies of Mesopotamia. J. Bombay Nat. Hist. S. 28: 50–70, 845–889.

Prater, S. H. 1920. The Arabian ostrich. J. Bombay Nat. Hist. S. 27: 602–605.

Rawi, Al-A. and H. L. Chakrawarthy. 1964. Medicinal plants of Iraq. Techn. Bull. Min. Agric. no. 15 Baghdad.

Rzóska, J. 1961. Observations on tropical rainpools and general remarks on temporary waters. Hydrobiologia, 17: 265–286.

Rzóska, J., ed. 1976. The Nile, Biology of an Ancient River. Junk, The Hague, Monographiae Biologicae 29.

Rzóska, J. 1978. On the Nature of Rivers. Junk, The Hague.

Saggs, H. W. F. 1962. The Greatness that was Babylon. A sketch of the ancient civilisations of the Tigris–Euphrates valley. Sedgwick & Jackson, London.

Scates, M. D. 1968. Notes on the hydrobiology of the Azraq Oasis, Jordan. Hydrobiologia, 1: 73–80.

Serruya, C. ed. 1978. Lake Kinneret. Monographiae Biologicae 32. Junk, The Hague.

Solecki, R. S. 1957. Two Neanderthal skeletons from Shanidar Cave. Sumer 13: 59–60.

Stark, Freya. 1951. Beyond Euphrates, Autobiography 1928–1933. J. Murray, London.

Stewart, T. D. 1958. First view of Shanidar Cave I skull. Sumer 14: 90–96.

Thalen, D. C. P. Ecology and Utilization of the desert-shrub rangelands of Iraq Script; Junk Publishers.

Thesiger, W. 1964. The Marsh Arabs. Longmanns, Green Co., London.

Ticehurst, C. B., P. A. Buxton, and R. E. Cheesman. 1922. The birds of Mesopotamia. J. Bombay Nat. Hist. S. 28: 210–250, 381–427, 650–674, 937–956.

Thompson, R. C. 1940. A Dictionary of Assyrian Botany. British Academy.

Troll, C. und Kh. Pappen. 1964. Karte der Jahreszeiten-Klimate der Erde. Erdkunde vol. 18. Reprod. also in Weltkarte zur Klimatologie. Springer Verlag Berlin etc. 1965.

Uvarov, B. P. 1922. Records and descriptions of Orthoptera from S.W. Asia. J. Bombay Nat. Hist. S. 28: 719–788.

Wiltshire, E. R. 1957. The Lepidoptera of Iraq. Rev. and enlarged edition. Dept. of Agric. Iraq, Bull. 30.

Woolley, C. L. 1950. Ur of the Chaldaees, A record of seven years of excavation. E. Brown, London, 2nd edition.

Woolley, Sir Leonard. 1960. Digging up the past. Pelican Books, Gt. Britain.

Worthington, E. B. 1946. Middle East Science. A survey of subjects other than agriculture. Report to the Middle East Supply Centre. H.M. Stationery Office, London.

Wright, H. E., Jr. 1955. Geologic aspects of the archaeology of Iraq. Sumer 11: 85–91.

Young, G. 1977. Return to the Marshes. Life with the Marsh Arabs of Iraq. Collins, London.

Zohari, M. 1971. Phytogeographical foundations of the Middle East. In: Davies et al., Plantlife of South-East Asia. Symposium, Tercentenary, Roy. Bot. Gardens Edinburgh.

Subject index

Together with the detailed Table of Contents the index should be useful for finding particular items of interest.